Alternative Energy

Trainee Guide

D0140611

Boston Columbus Indianapolis New York San Francisco Amsterdam
Cape Town Dubai London Madrid Milan Munich Paris Montreal Toronto
Delhi Mexico City São Paulo Sydney Hong Kong Seoul Singapore Taipei Tokyo

National Center for Construction Education and Research
President: Don Whyte
Director of Product Development: Daniele Dixon
Alternative Energy Project Manager: Matt Tischler
Production Manager: Tim Davis

Quality Assurance Coordinator: Debie Ness
Desktop Publishing Coordinator: James McKay
Production Specialist: Heather Griffith-Gatson
Editor: Chris Wilson

Writing and development services provided by Topaz Publications, Liverpool, NY
Lead Writer/Project Manager: Troy Staton
Desktop Publisher: Joanne Hart
Art Director: Megan Paye

Permissions Editors: Andrea LaBarge, Alison Richmond
Writers: Troy Staton, Pat Vidler, Veronica Westfall

Pearson Education, Inc.
Head of Global Certifications/Associations: Andrew Taylor
Editorial Assistant: Collin Lamothe
Program Manager: Alexandrina B. Wolf
Digital Studio Project Managers: Heather Darby,
  Tanika Henderson, Jose Carchi

Director of Marketing: Leigh Ann Simms
Senior Marketing Manager: Brian Hoehl
Cover photo: ©istockphoto.com

Composition: NCCER
Printer/Binder: LSC Communications
Cover Printer: LSC Communications
Text Fonts: Palatino and Univers

Copyright © 2011 by NCCER, Alachua, FL 32615, and published by Pearson Education, Inc., New York, NY 10013. All rights reserved. Printed in the United States of America. This publication is protected by Copyright and permission should be obtained from NCCER prior to any prohibited reproduction, storage in a retrieval system, or transmission in any form or by any means, electronic, mechanical, photocopying, recording, or likewise. For information regarding permission(s), write to: NCCER Product Development, 13614 Progress Blvd., Alachua, FL 32615.

PEARSON

www.pearsonhighered.com

ISBN-13: 978-0-13-266625-1
ISBN-10:   0-13-266625-1

Preface

To the Trainee

As the demand for energy grows, the need to supplement current fossil fuel sources also increases. Recent technological advancements have allowed for the use of existing and plentiful alternative fuel supplies, such as biomass, wind power, solar energy, and nuclear power.

NCCER's *Alternative Energy* curriculum covers the production, associated technology, and uses of all of these types of unconventional fuel sources. In addition to containing the latest information about each energy source, every module includes interesting and thought-provoking activities about the function and future of alternative energy.

Alternative Energy is endorsed by the Florida Energy Workforce Consortium (FEWC), and this title supports the Florida Department of Education's Curriculum Frameworks for Energy Technician, Power Distribution Technician, and Energy Generation Technician. In addition, *Alternative Energy* supports Tier 3 and Tier 4 of the Get into Energy Competency Model as outlined by the Center for Energy Workforce Development (CEWD) and the Employment and Training Administration (ETA).

Skilled workers are always in demand in the energy industry. Candidates with diversified backgrounds in both fossil and alternative fuels will find successful and gratifying opportunities in this expanding field. NCCER's *Alternative Energy* training program provides trainees with the knowledge required to advance their careers in this challenging and rewarding sector.

We wish you success as you progress through this training program. Should you have any comments on how NCCER might improve upon this textbook, please complete the User Update form located at the back of each module and send it to us. We will always consider and respond to input from our customers.

We invite you to visit NCCER's website at **www.nccer.org** for information on the latest product releases and training, as well as online versions of the *Cornerstone* newsletter and Pearson's product catalog.

Your feedback is welcome. You may email your comments to **curriculum@nccer.org** or send general comments and inquiries to **info@nccer.org**.

NCCER Curricula

NCCER is a not-for-profit 501(c)(3) education foundation established in 1996 by the world's largest and most progressive construction companies and national construction associations. It was founded to address the severe workforce shortage facing the industry and to develop a standardized training process and curricula. Today, NCCER is supported by hundreds of leading construction and maintenance companies, manufacturers, and national associations. The NCCER Standardized Curricula was developed by NCCER in partnership with Pearson, the world's largest educational publisher.

Some features of the NCCER Standardized Curricula are as follows:

- An industry-proven record of success
- Curricula developed by the industry for the industry
- National standardization providing portability of learned job skills and educational credits
- Compliance with the Office of Apprenticeship requirements for related classroom training (*CFR 29:29*)
- Well-illustrated, up-to-date, and practical information

NCCER also maintains a National Registry that provides transcripts, certificates, and wallet cards to individuals who have successfully completed a level of training within a craft in NCCER's Curricula. *Training programs must be delivered by an NCCER Accredited Training Sponsor in order to receive these credentials.*

Special Features

In an effort to provide a comprehensive, user-friendly training resource, we have incorporated many different features for your use. Whether you are a visual or hands-on learner, this book will provide you with the proper tools to get started in the field of alternative energy.

Introduction

This page is found at the beginning of each module and lists the Objectives and Trade Terms for that module. The Objectives list the skills and knowledge you will need in order to complete the module successfully. The list of Trade Terms identifies important terms you will need to know by the end of the module.

74104-11
SOLAR POWER

Objectives

When you have completed this module, you will be able to do the following:

1. Define solar power, how it is harnessed, and how it is used to generate energy.
2. List the advantages and disadvantages of solar energy.
3. Describe the past, present, and future of solar energy.
4. Identify and describe solar applications.

Performance Tasks

This is a knowledge-based module; there are no performance tasks.

Trade Terms

Air mass
Altitude
Ambient temperature
Amorphous
Array
Autonomy
Azimuth
Backfeed
Balance of system (BOS)
Brownout
Building-integrated photovoltaics (BIPV)
Charge controller
Combiner box
Concentrating collector
Declination
Depth of discharge (DOD)
Doped
Dual-axis tracking
Electrochemical solar cell
Elevation
Evacuated tube collector
Flat plate collector
Fuel cells
Grid-connected system
Grid-interactive system
Grid-tied system
Heliostats
Hybrid system

Insolation
Integral-collector storage system
Inverter
Irradiance
Latitude
Maximum power point tracking (MPPT)
Module
Monocrystalline
Net metering
Off-grid system
Peak sun hours
Polycrystalline
Pulse width-modulated (PWM)
Sea level
Semiconductor
Sine wave
Single-axis tracking
Solar photovoltaic (PV) system
Solar thermal system
Spectral distribution
Standalone system
Standard Test Conditions (STC)
Sun path
Thermosiphon system
Thin film
Tilt angle
Utility-scale solar generating system
Watt-hours (Wh)

Industry Recognized Credentials

If you're training through an NCCER-accredited sponsor you may be eligible for credentials from NCCER's Registry. The ID number for this module is 74104-11. Note that this module may have been used in other NCCER curricula and may apply to other level completions. Contact NCCER's Registry at 888.622.3720 or go to nccer.org for more information.

Color Illustrations and Photographs

Full-color illustrations and photographs are used throughout each module to provide vivid detail. These figures highlight important concepts from the text and provide clarity for complex instructions. Each figure reference is denoted in the text in *italic type* for easy reference.

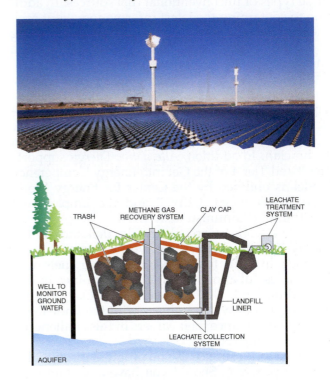

Notes, Cautions, and Warnings

Safety features are set off from the main text in highlighted boxes and are organized into three categories based on the potential danger of the issue being addressed. Notes simply provide additional information on the topic area. Cautions alert you of a danger that does not present potential injury but may cause damage to equipment. Warnings stress a potentially dangerous situation that may cause injury to you or a co-worker.

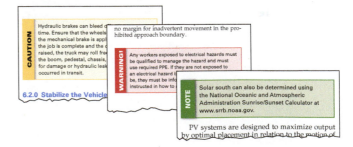

Did You Know?

The Did You Know? features offer hints, tips, and other helpful bits of information from the trade.

Did You Know?

Sugarcane is a Cool Crop

Brazil's 30-year-old sugarcane-to-ethanol project has been a success. Sugarcane fuels almost 25 percent of the country's automobiles. Research now shows another benefit—sugarcane cools the air where it grows. Its ability to reflect sunlight, draw cooling moisture from the soil, and release it into the air helps to lower the air temperature surrounding it. Carnegie Institution's Department of Global Ecology researchers say that when a sugarcane crop replaces other crops, the local climate cools.

less carbon monoxide than gasoline does. They can be made from local crops, eliminating the ... to a lesser degree; however, it has less effect on mileage than E85, a compro...

On Site

On Site features provide a head start for those entering the alternative energy fields by presenting technical tips and professional practices from energy workers on a variety of topics. On Site features often include real-life scenarios similar to those you might encounter on the job site.

On Site

Current Research

At North Carolina State University, in cooperation with state and federal institutions, private companies, and other universities, research is taking place to analyze the best biomass for fuel and the environment. The goal is a crop that can be harvested all year instead of switchgrass and sorghum, which have three- to five-month harvest times.

In Florida, two little-known crops are being grown between crops or on fallow land. Camelina is grown for its oily seed. It can be used for both aviation fuel and animal food. The second crop is kenaf (shown in the figure). It is fast-growing biomass used mainly for oil and ethanol.

5.3.1 Research and Development

Think About It

Think About It features use "What if?" questions to help you apply theory to real-world experiences and put your ideas into action.

Think About It

The Waste We Produce

Americans generate more than 1,600 pounds of waste a year for each person. In addition, today's lifestyle uses huge amounts of disposable material in the form of plastic, Styrofoam, and paper. Each person's garbage could take up two cubic yards of landfill a year. That is about the size of a large refrigerator box. The population of the US is well over 300 million. That much garbage takes a great deal of space.

Step-by-Step Instructions

Step-by-step instructions are used throughout to guide you through technical procedures and tasks from start to finish. These steps show you not only how to perform a task but also how to do it safely and efficiently.

trically safe work condition. Most companies have detailed written procedures for performing this task.

Step 1 Determine whether any other crews are working on the circuit. All distribution lines are treated as if they are energized unless your team has performed a de-energizing procedure. If two or more crews are working on the same lines or equipment, each crew must independently perform a de-energizing procedure and apply their own lockout/tagout devices to energy controls.

Step 2 Designate one qualified member of the crew as the employee in charge of the electrical clearance.

Trade Terms

Each module presents a list of Trade Terms that are discussed within the text and defined in the Glossary at the end of the module. These terms are denoted in the text with **blue, bold type** upon their first occurrence. To make searches for key information easier, a comprehensive Glossary of Trade Terms from all modules is located at the back of this book.

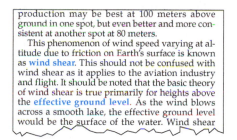

production may be best at 100 meters above ground in one spot, but even better and more consistent at another spot at 80 meters.

This phenomenon of wind speed varying at altitude due to friction on Earth's surface is known as wind shear. This should not be confused with wind shear as it applies to the aviation industry and flight. It should be noted that the basic theory of wind shear is true primarily for heights above the effective ground level. As the wind blows across a smooth lake, the effective ground level would be the surface of the water. Wind shear

Review Questions

Review Questions are provided to reinforce the knowledge you have gained. This makes them a useful tool for measuring what you have learned.

Review Questions

1. Assuming all conditions remain exactly the same, about how many years of known global oil reserves remain at present rate of use?
   a. 15
   b. 25
   c. 40
   d. 60

2. All alternative energy sources are also considered renewable.
   a. True
   b. False

3. How much electrical power does an average American household consume every hour?
   a. 1 kW
   b. 1 MW
   c. 100 watts
   d. 8,500 watts

7. Of the electrical power that is produced without carbon emissions, nuclear power contributes _____.
   a. 20 percent
   b. 45 percent
   c. 70 percent
   d. 85 percent

8. The primary reason for the limited future expansion of hydroelectric power in the United States is because _____.
   a. the facilities cost too much to operate and maintain
   b. the environmental damage is considered too great
   c. most of the ideal locations have already been developed
   d. it does not provide a significant amount of power per facility

NCCER Curricula

NCCER's training programs comprise more than 80 construction, maintenance, pipeline, and utility areas and include skills assessments, safety training, and management education.

Boilermaking
Cabinetmaking
Carpentry
Concrete Finishing
Construction Craft Laborer
Construction Technology
Core Curriculum:
    Introductory Craft Skills
Drywall
Electrical
Electronic Systems Technician
Heating, Ventilating, and
    Air Conditioning
Heavy Equipment Operations
Highway/Heavy Construction
Hydroblasting
Industrial Coating and Lining
    Application Specialist
Industrial Maintenance
    Electrical and Instrumentation
    Technician
Industrial Maintenance
    Mechanic
Instrumentation
Insulating
Ironworking
Masonry
Millwright
Mobile Crane Operations
Painting
Painting, Industrial
Pipefitting
Pipelayer
Plumbing
Reinforcing Ironwork
Rigging
Scaffolding
Sheet Metal
Signal Person
Site Layout
Sprinkler Fitting
Tower Crane Operator
Welding

Green/Sustainable Construction

Building Auditor
Fundamentals of
    Weatherization
Introduction to Weatherization
Sustainable Construction
    Supervisor
Weatherization Crew Chief
Weatherization Technician
Your Role in the Green
    Environment

Energy

Alternative Energy
Introduction to the Power
    Industry
Introduction to Solar
    Photovoltaics
Introduction to Wind Energy
Power Industry Fundamentals
Power Generation Maintenance
    Electrician
Power Generation I&C
    Maintenance Technician
Power Generation Maintenance
    Mechanic
Power Line Worker
Solar Photovoltaic Systems
    Installer
Wind Turbine Maintenance
    Technician

Pipeline

Control Center Operations,
    Liquid
Corrosion Control
Electrical and Instrumentation
Field Operations, Liquid
Field Operations, Gas
Maintenance
Mechanical

Safety

Field Safety
Safety Orientation
Safety Technology

Management

Fundamentals of Crew
    Leadership
Project Management
Project Supervision

Supplemental Titles

Applied Construction Math
Careers in Construction
Tools for Success

Spanish Translations

Basic Rigging
    (Principios Básicos de
    Maniobras)
Carpentry Fundamentals
    (Introducción a la Carpintería,
    Nivel Uno)
Carpentry Forms
    (Formas para Carpintería, Nivel
    Trés)
Concrete Finishing, Level One
    (Acabado de Concreto, Nivel
    Uno)
Core Curriculum:
    Introductory Craft Skills
    (Currículo Básico: Habilidades
    Introductorias del Oficio)
Drywall, Level One
    (Paneles de Yeso, Nivel Uno)
Electrical, Level One
    (Electricidad, Nivel Uno)
Field Safety
    (Seguridad de Campo)
Insulating, Level One
    (Aislamiento, Nivel Uno)
Ironworking, Level One
    (Herrería, Nivel Uno)
Masonry, Level One
    (Albañilería, Nivel Uno)
Pipefitting, Level One
    (Instalación de Tubería
    Industrial, Nivel Uno)
Reinforcing Ironwork, Level One
    (Herreria de Refuerzo, Nivel
    Uno)
Safety Orientation
    (Orientación de Seguridad)
Scaffolding
    (Andamios)
Sprinkler Fitting, Level One
    (Instalación de Rociadores,
    Nivel Uno)

Acknowledgments

This curriculum was revised as a result of the farsightedness and leadership of the following sponsors:

Florida Energy Workforce Consortium
Gulf Power
Lakeland Electric
Madison Comprehensive High School
Tri-City Electrical Contractors, Inc.
Vision Quest Academy
Xcel Energy

This curriculum would not exist were it not for the dedication and unselfish energy of those volunteers who served on the Authoring Team. A sincere thanks is extended to the following:

Tanzania Adams
Tim Dean
Sybelle Fitzgerald
Jennifer Grove
Betsy Levingston

Lonnie Noack
Kent Peterson
Mike Powers
Gordon Johnson
Tony Vasquez

Florida Energy Workforce Consortium Members

JEA	Gulf Power
CHELCO	FP&L
Lakeland Electric	Tampa Electric
OUC	Company
Progress Energy	

FEWC Workforce Partners

Workforce Florida, Inc.
CLM Workforce Connection

NCCER Partners

American Council for Construction Education
American Fire Sprinkler Association
Associated Builders and Contractors, Inc.
Associated General Contractors of America
Association for Career and Technical Education
Association for Skilled and Technical Sciences
Construction Industry Institute
Construction Users Roundtable
Design Build Institute of America
GSSC – Gulf States Shipbuilders Consortium
ISN
Manufacturing Institute
Mason Contractors Association of America
Merit Contractors Association of Canada
NACE International
National Association of Women in Construction
National Insulation Association
National Technical Honor Society
NAWIC Education Foundation
North American Crane Bureau
North American Technician Excellence

Pearson
Prov
SkillsUSA®
Steel Erectors Association of America
U.S. Army Corps of Engineers
University of Florida, M. E. Rinker Sr., School of Construction Management
Women Construction Owners & Executives, USA

NCCER Business Partners

Contents

Module One

Introduction to Alternative Energy

Identifies the need for alternative energy development. Describes the contributions and potential of individual alternative energy sources. Also covers the present US electrical grid and issues affecting specific alternative energy source tie-in and reliability. (Module ID number 74101-11; 25 Hours)

Module Two

Biomass and Biofuels

Defines potential sources of biomass and biofuels and discusses their advantages and disadvantages for energy production. Discusses the future of biomass as well as biomass energy applications. (Module ID number 74102-11; 22.5 Hours)

Module Three

Nuclear Power

Describes nuclear power and its sources. Discusses the advantages and disadvantages of nuclear power, the future of nuclear energy, and nuclear power generation. (Module ID number 74103-11; 25 Hours)

Module Four

Solar Power

Describes solar photovoltaic (PV) power and how it is harnessed. Identifies the advantages and disadvantages of solar energy. Discusses the past, present, and future of solar energy, as well as solar PV applications. (Module ID number 74104-11; 25 Hours)

Note: *NFPA 70®*, *National Electrical Code®*, and *NEC®* are registered trademarks of the National Fire Protection Association, Inc., Quincy, MA 02269. All *National Electrical Code®* and *NEC®* references in this module refer to the 2011 edition of the *National Electrical Code®*.

Module Five

Wind Power

Describes wind power and how it is harnessed. Identifies the advantages and disadvantages of wind energy. Discusses the past, present, and future of wind energy, as well as wind energy applications. (Module ID number 74105-11; 22.5 Hours)

Glossary

Index

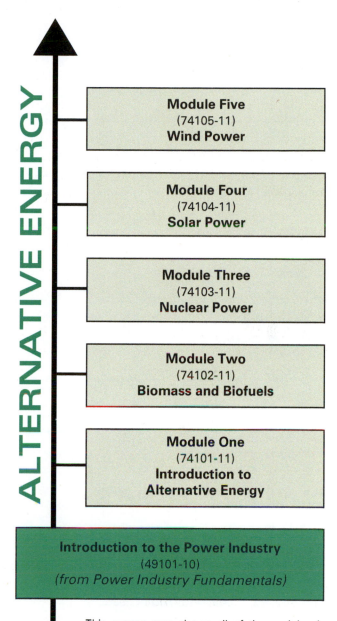

This course map shows all of the modules in *Alternative Energy*. The suggested training order begins at the bottom and proceeds up. Skill levels increase as you advance on the course map. The local Training Program Sponsor may adjust the training order.

74101-11

Introduction to Alternative Energy

Module One

Trainees with successful module completions may be eligible for credentialing through NCCER's National Registry. To learn more, go to **www.nccer.org** or contact us at **1.888.622.3720**. Our website has information on the latest product releases and training, as well as online versions of our *Cornerstone* newsletter and Pearson's product catalog.

Your feedback is welcome. You may email your comments to **curriculum@nccer.org,** send general comments and inquiries to **info@nccer.org**, or use the User Update form at the back of this module.

Copyright © 2011 by the National Center for Construction Education and Research (NCCER) and published by Pearson Education, Inc., publishing as Prentice Hall. All rights reserved. Manufactured in the United States of America. This publication is protected by Copyright, and permission should be obtained from NCCER prior to any prohibited reproduction, storage in a retrieval system, or transmission in any form or by any means, electronic, mechanical, photocopying, recording, or likewise. To obtain permission(s) to use material from this work, please submit a written request to NCCER Product Development, 13614 Progress Blvd., Alachua, FL 32615.

Objectives

When you have completed this module, you will be able to do the following:

1. Understand the need for alternative energy and identify the various forms.
2. Describe the contributions of alternative energy sources to world supplies at present and their potential.
3. Describe the present US electrical grid and issues affecting alternative energy source tie-in, reliability, and economic impact.

Performance Tasks

This is a knowledge-based module; there are no performance tasks.

Trade Terms

Alternative energy
Base load unit
Biofuel
Biomass
Centralized power generation
Cogeneration
Combined heat and power (CHP)
Concentrated solar thermal (CST)
Electric grid
Fuel cells
Gasification
Greenhouse gases
Gross Domestic Product (GDP)
Interconnection agreement
Kilowatt

Kilowatt-hour (kWh)
Megawatt (MW)
Microgrid
Net metering
Peaking load unit
Phantom loads
Photovoltaic (PV)
Renewable energy
Renewable Portfolio Standard (RPS)
Smart grid
Solar thermal energy (STE)
Superconductors
Utility-scale
Watt
Zero energy district

Industry Recognized Credentials

If you're training through an NCCER-accredited sponsor you may be eligible for credentials from NCCER's Registry. The ID number for this module is 74101-11. Note that this module may have been used in other NCCER curricula and may apply to other level completions. Contact NCCER's Registry at 888.622.3720 or go to nccer.org for more information.

Contents

Topics to be presented in this module include:

Figures and Tables

1.0.0 INTRODUCTION

Fossil fuels (coal, oil, and natural gas) provide roughly 85 percent of all US energy needs. Almost all transportation fuels and nearly two-thirds of US electrical power are generated from a fossil-fuel source. It is safe to say that fossil fuels have helped propel the world into the twenty-first century, and the world would be a very different place today without them. Fossil fuels remain an enormous part of daily life. However, there are issues related to their use that must be considered and addressed.

The issue begins with the supply of fossil fuels. If there are absolutely no changes in previous patterns of usage or technology, some private industry estimates believe that global oil reserves could be used up in about 40 years. Under the same circumstances, the supply of natural gas may last about 60 years. Such estimates vary widely though, and recent technology that locates and extracts natural gas from shale formations extends the estimate for natural gas up to 110 years, according to the Energy Information Administration (EIA). Considering how long it will take to change energy behavior and put new energy sources to work, these are very short periods of time. However, new technology may lead to the discovery of new oil and gas reserves, and help access those previously thought to be inaccessible. There are many variables that will influence the actual life of fossil-fuel supplies.

Meanwhile, American access to the proven resources that remain is worthy of consideration. The twelve member countries of OPEC (Organization of Petroleum Exporting Countries, an international governmental trade group) control nearly 80 percent of the proven oil resources, according to the EIA. Several countries own a great deal of the world's proven natural gas reserves. There is a great deal of distance, both physical and political, between the fossil-fuel resources and the points of greatest consumption. However, it must also be noted that it is difficult to identify and quantify resources yet to be discovered. There may be more accessible fossil fuel than anyone realizes.

The second part of the issue is related to the environment. The use of fossil fuel releases greenhouse gases to the atmosphere. Greenhouse gases act like an umbrella around the earth, preventing heat from leaving. Some greenhouse gases occur naturally, and the Environmental Protection Agency (EPA) suggests the planet would be roughly 60°F cooler if none existed at all. Many individuals believe that the planet is measurably warming, and that the additional greenhouse gases in the atmosphere resulting from human activity are the reason. Other individuals believe continued increases could someday result in catastrophic melting of the polar ice caps. According to the EPA's website, increased greenhouse gases are "likely contributing to an increase in global average temperature and related climate changes."

The main greenhouse gas is carbon dioxide, or CO_2. The EPA estimates that 95 percent of the CO_2 emissions come from the use of fossil fuels. Note that this figure is for all purposes, not just electric power generation alone. Beyond atmospheric issues, the retrieval of fossil fuels from the earth can also have significant environmental effects.

Many of today's efforts to develop new energy sources began in 1973 with the first US oil embargo. The ensuing energy crisis got everyone's attention. Since that time, the federal government has invested well over $100 billion in alternative energy and efficiency research. Today, according to the EIA, the United States still accounts for 25 percent of the greenhouse gas emitted globally. This is primarily due to the United States having the largest economy in the world and 85 percent of its energy needs are provided by fossil fuels. A 2010 report from the International Energy Agency (IEA) indicates that, as of 2008, only China exceeded the volume of US CO_2 emissions.

The US figures seem significantly better when emissions are compared to the Gross Domestic Product (GDP) of various countries. Industrialized countries naturally have higher emissions. When related to the GDP, US emissions of CO_2 fell nearly 5 percent between 2007 and 2008, while China's emissions rose by 8 percent. Although efforts to reduce emissions are helping, there is clearly a need to continue the endeavor globally. By developing clean energy technology, the United States can fill a global need for such technology, helping to protect the environment while prospering economically.

In spite of efforts to develop alternative energy sources quickly, demand for power and the use of fossil fuels continues to rise. Although shrinking fossil-fuel resources and environmental effects are on the minds of humans globally, consistent, practical, and economically sound efforts must be made before significant change is realized.

Fortunately, the United States has organizations like the National Renewable Energy Laboratory (NREL) (*Figure 1*) to provide the needed energy research and guidance going forward. This module will explore the work of such agencies and alternative and renewable energy resources; the realistic potential they represent to offset the use of fossil fuels in the United States; and the issues affecting their use. The following energy sources will be considered:

- Biomass and biofuels
- Nuclear
- Solar
- Wind
- Hydroelectric, geothermal, and other practical sources

Each of these sources will be reviewed, along with some specific details related to their present and future use.

2.0.0 ALTERNATIVE ENERGY OVERVIEW

The terms *alternative* and *renewable* are often used interchangeably, but they are not the same. Alternative energy can be defined as any source of energy that provides an alternative to a fossil-fuel source. In general, alternative energy sources are needed to provide electric power generation without the concerns related to the use of fossil fuels. They must also be a functional replacement, providing comparable service to fossil fuels. However, as is true of the subject as a whole, what qualifies as alternative energy is often debated.

Renewable energy can be defined as an energy source that is naturally replenished (*Figure 2*). Fossil fuels are created by natural means and they need only be accessed. However, the creation process is painfully slow. Access to some sources is challenging or impossible. Without question, fossil fuels are consumed faster than they can re-

generate, and the possibility exists that the supply will become unreliable. The business and economics of fossil fuels is already complex. As the resources become more limited, the control and distribution strategies of those resources will become more complex as well.

Most alternative energy sources can also be defined as renewable. The sun, wind, and water are likely the first sources that come to mind. Most biomass is typically defined as renewable. Although the term *biomass* may be relatively new, its use as a source of heat and energy is well documented throughout history. Nuclear energy, though, is typically not considered renewable. Some argue that uranium resources are unlimited, and therefore can be considered renewable. Others say the present rate of use would eventually consume the supply. The debate will likely continue on this subject, but it is somewhat irrelevant. Nuclear power is already in widespread use, and it represents a practical, no-emission alternative energy source that can last for many years.

Figure 3 provides a visual representation of the contribution of various energy sources in the United States for 2009, according to the EIA. Note that these specific figures are related to all energy consumed, including transportation and heating fuels. Source contributions related specifically to electric power generation are examined later in this module. Renewable energy sources provided about 8 percent of the energy, while nuclear energy supplied about 9 percent of the total. The

Figure 1 NREL visitors center.

Figure 2 Renewable energy sources.

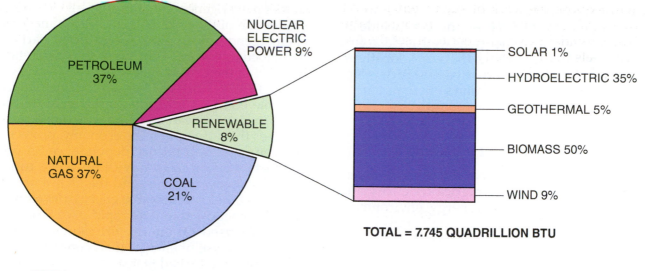

TOTAL = 94.820 QUADRILLION BTU

TOTAL = 7.745 QUADRILLION BTU

74101-11_F03.EPS

Figure 3 National energy consumption by source, 2009.

total then, from all alternative energy sources, is about 17 percent. Although only about 1 percent of electric power generation comes from petroleum, 37 percent of the total energy consumed in the United States originates from it. The most significant use of petroleum in the United States is related to transportation.

2.1.0 National Position and Support

The United States has a diverse set of energy resources and the innovative skills required to make the most of them. A review of the government's position on energy and the federal agencies that manage those resources is vital to understanding the role you can play in the future.

2.1.1 Energy Independence and Security Act (EISA)

The Energy Independence and Security Act (EISA) of 2007 established energy and fossil fuel dependence reduction guidelines for federal buildings and facilities. The goals of the EISA include the following:

- Reducing the energy consumption of facilities by 3 percent each year, or by 30 percent by 2015
- Implementing renewable energy generation projects on site of various facilities
- Reducing the use of fossil fuels in new and renovated buildings by 55 percent in 2010, increasing to 100 percent by 2030
- Generate 30 percent of hot water from solar energy in new and renovated buildings

The Act includes far more than these goals, but it is evident that the federal government has taken

the adoption of energy management and alternative energy very seriously. The Department of Defense (DoD) is working toward purchasing or generating 7.5 percent of its energy needs from renewable sources by 2013. By 2025, it hopes to raise that figure to 25 percent, according to the *Department of Defense Facilities and Vehicles Energy Use, Strategies and Goals Program Overview*, dated May 11, 2009.

2.1.2 Blueprint for a Secure Energy Future

In 2011, the White House released its *Blueprint for a Secure Energy Future*. The plan outlines the following three primary strategies:

- Develop and secure America's energy supplies. This includes the development of domestic oil and gas supplies in a safe and responsible manner. In 2010, domestic crude oil production reached its highest level since 2003.
- Provide consumers with choices to reduce costs and save energy. Since transportation is responsible for more than 70 percent of petroleum consumption, greater fuel efficiency for all vehicles will be a continued point of focus. In addition, the administration has set a goal of having 1,000,000 electric vehicles on American roads by 2015. Increasing the efficiency of the American home through better construction approaches and weatherization is also a priority.
- Innovate a way to a clean energy future. The administration acknowledges that a global race is under way to develop clean energy technology, and that the United States must be in the race to win. To that end, the President has proposed to generate 80 percent of the country's

electrical power from a diverse set of clean energy sources by 2035. Those sources include all forms of alternative energy, as well as clean coal and natural gas. Leading the way globally in clean fossil fuel and alternative energy development will enhance the nation's energy security and help secure its economic future.

2.1.3 The US Department of Energy

The US Department of Energy (DOE) reports that its mission is to "ensure America's security and prosperity by addressing its energy, environmental, and nuclear challenges through transformative science and technology solutions." One focus is on advancements in science and technology that result in a better quality of life overall.

The DOE recognizes the importance of alternative energy to the future and is a steadfast supporter of its use. It also recognizes that, with 85 percent of US energy originating from fossil fuels now, the advancement of alternate sources is of the highest priority. There are many obstacles to integrating alternate sources into the national power supply, but the demand for more electric power is relentless. As alternative sources are more fully developed, wiser use of fossil fuels cannot be ignored.

The DOE's Office of Fossil Energy's (FE) primary mission is "to ensure the nation can continue to rely on traditional resources for clean, affordable energy while enhancing environmental protection." The FE's staff includes roughly 1,000 scientists, engineers, technicians, and administrative staff. While others focus on the alternatives, FE focuses on the reduction of emissions associated with fossil fuel power generation; the efficient capture and production of the resources; and the management of emergency oil supplies. A significant part of its work focuses on coal—the most abundant energy resource in the United States.

The DOE's Office of Energy Efficiency and Renewable Energy (EERE) has a different mission. EERE invests in clean and renewable energy technologies that strengthen the economy, protect the environment, and reduce dependence on oil. It conducts work in unison with business, universities, local governments, and the DOE's national labs. EERE helps renewable energy develop and become practical at a more rapid pace. The office is also focused on energy efficiency, supporting the wise use of energy at all consumer levels throughout the United States.

The National Renewable Energy Laboratory operates under the guidance of the EERE office, but it also receives funding from other sources outside of the government. The NREL, headquartered near Boulder, CO, is the primary laboratory for renewable energy and energy efficiency testing. A variety of testing is conducted there and at various satellite locations. Research and development conducted by the NREL helps to advance energy goals. It provides private industry and the energy market with technology and testing platforms that might otherwise be unavailable.

One DOE office, the Office of Nuclear Energy (NE), promotes the use of nuclear power. It supports the technology to meet energy, environmental, and national security needs. Due to the unique security and management precautions required, a separate office to manage nuclear programs is a necessity.

Solid-State Lighting

GOING GREEN

According to the DOE's Office of Energy Efficiency and Renewable Energy, solid-state lighting (SSL) offers more potential than any other lighting technology to save energy and enhance the quality of building environments, contributing to the nation's energy and climate change solutions. Research and development, based primarily on the light-emitting diode (LED), continues to improve SSL performance and application. One distinct advantage is the absence of mercury, a hazardous component of compact fluorescent lamps (CFL). The functional life of SSLs is also very impressive.

74101-11_SA01.EPS

Another DOE office plays a key role in the use of alternative energy. The Office of Electricity Delivery and Energy Reliability (OE) seeks to improve and protect the security and reliability of the entire power system; modernize and enhance the national electric grid; and assist in the recovery from disruptions to the power supply. The generation of power from alternative energy sources is only one part of energy strategy. In order to make use of that energy, new methods of controlling the flow of power from the point of generation to the consumer are necessary.

These offices must all work together to integrate alternative energy into the energy mix. In addition to the work of these offices, federal and state governments provide incentives to individuals willing to invest in alternative energy research and use. Through tax credits, consumers who install alternative energy systems can offset a significant part of the cost. Once installed, many users also enjoy the benefit of selling any excess power back to the utility. This could provide an additional return on their investment.

2.2.0 State Programs

Each state of the United States differs to some degree in their resources and needs, including their available renewable energy resources. As a result, it makes sense to allow states an opportunity to develop some individual goals regarding energy.

One tool for states is known as the **Renewable Portfolio Standard (RPS)**. An RPS typically places responsibilities on electric utilities to supply a specific portion of consumed power from renewable energy sources. The goal is to encourage and inspire the development of renewable sources, based on the idea that a greater volume of available renewable energy producing resources will bring their generating cost in line with that of traditional sources. An RPS can also include the use of **combined heat and power (CHP)** approaches. Through an RPS, state governments can guide the power industry in the direction they choose. Additional energy efficient practices may be instituted in conjunction with an RPS as a means of reducing demand and consumption.

States also provide incentives for the use of renewable energy sources and energy efficient devices. The Database of State Incentives for Renewables & Efficiency (DSIRE) is a program initiated by the North Carolina Solar Center and the Interstate Renewable Energy Council (IREC). It is funded by the DOE's EERE office and administered by the NREL. As the name implies, the database is a comprehensive list of financial incentives offered in each state. A wide array of information is made available, and it covers residential, commercial, and utility operations. The number of incentives offered is quite staggering. Although states are the focus, federal incentives are listed as well.

2.3.0 Energy Efficiency

It has been said by many people, "the cleanest megawatt (MW) is the one never generated." Although energy efficiency is not related directly to alternative or renewable energy, some discussion of the issue is certainly relevant. One reason alternative strategies are needed is the continuous growth in the demand for power. Using the power generated, regardless of the source, in an efficient and conservative manner reduces the overall demand. While the strategy and technology to use alternative energy sources continues to develop, every effort must be made to use electrical power wisely. Energy efficiency is a prudent strategy, no matter how the power is generated.

The relationship between energy efficiency and new forms of energy production is strengthened by the EERE. It supports and invests in many forms of renewable energy research. It also provides the necessary information and guid-

---

On Site

ARPA-E

One little-known agency operating under DOE control is the Advanced Research Projects Agency – Energy (ARPA-E). The office was created in 2007 as a means of funding high-risk, high-reward research in the field of energy. The projects considered and funded are those that are too risky for private sources of funding to consider. Research on existing forms of energy is not considered at any level; other agencies support such projects.

It is hoped that ARPA-E funded projects will reveal scientific discoveries and inventions that will position the United States ahead of the world in energy research. In September 2010, six amazing projects were chosen from thousands of applicants to receive the latest round of grants. One example project is the development of airborne wind turbines. Each project has the potential to change forever how we generate power or use energy.

---

ance to government operations and the public to manage energy use effectively. It is essential that all energy be used wisely to help offset the need for additional power generation. Energy-efficient practices are becoming increasingly important. Everyone can contribute to the cause. The average American household consumes energy as shown in *Figure 4*, per the EERE's Energy Savers website. Most of the energy goes to power heating and cooling systems, hot water heating, and common appliances.

One surprising fact is the 8 percent energy consumption by other devices. Standby power is part of this category. Standby power loads are electrical devices that consume energy when they are not being actively used, and do so unless they are physically disconnected from a power source. They are sometimes referred to as phantom loads. However, phantom loads typically refer to devices that have no real purpose to remain powered. For example, a security system would be considered a standby power load, but not a phantom load since it is performing the task of monitoring the home. A TV turned off would be considered both a standby power load and a phantom load. *Figure 5* provides some examples of standby power loads and the amount of energy they consume, based on testing done at the Lawrence Berkeley National Laboratory. Individually, the loads may seem insignificant. However, the combined load has a substantial impact on total consumption. The DOE, according to its Energy Star website, estimates that 75 percent of the electricity used to power home electronics and appliances is consumed while the products are turned off. In order

for energy efficiency to help reduce the hunger for power, manufacturers as well as consumers must embrace the need to reduce standby power consumption. Both state and federal agencies encourage energy retrofits and the use of high-efficiency appliances at all consumer levels. This is done through tax incentives and sharing valuable data on the subject.

On the commercial and business side, one relatively new and growing electrical appetite is related to the Internet and the many server farms that support it and the storage of data. In separate 2006 magazine articles, tech specialists Stephanie Mehta and Ron Starner estimate that a single click to search an item on the Internet activated roughly 7,000 computers to participate. Daniel McDonell, a specialist in sustainable development, estimates that a modestly sized server farm may use 15 MW of power each day, including the necessary cooling systems. A single megawatt of power is sufficient to power roughly 250 average homes. The demands of technology obviously require a significant amount of power, and server farms to support Internet-related activity continue to rise. Advancing the energy-efficient use of technical resources such as server farms could potentially save tremendous amounts of electrical power.

2.4.0 Obstacles to Alternative Energy Development

One significant obstacle to more rapid alternative energy use is certainly technology. Technology simply has not advanced far enough yet to allow us to replace fossil fuels overnight. Perhaps it would have been possible by now if the interest in alternative energy development had started many years earlier. The world of energy, from the extraction of raw materials to power production, primarily relies on fossil fuels. Not only do they provide the energy to generate electrical power, they also fuel the vehicles and equipment required to extract and transport them. Alternative energy forms, then, are challenged to replace both fossil-fuel power sources as well as the fuels of the supply chain. It is essential that research and development efforts continue to allow technology to catch up with needs and provide the answers. Each form of alternative energy has its own set of technological challenges to its widespread use.

Cost is another significant factor that inhibits the greater use of alternative energy. If the technology were available, could we afford it? The question of cost will continue, even if all the answers are provided by technology. A day may come

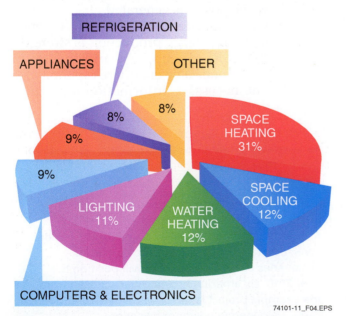

74101-11_F04.EPS

Figure 4 Energy consumption in the average US household.

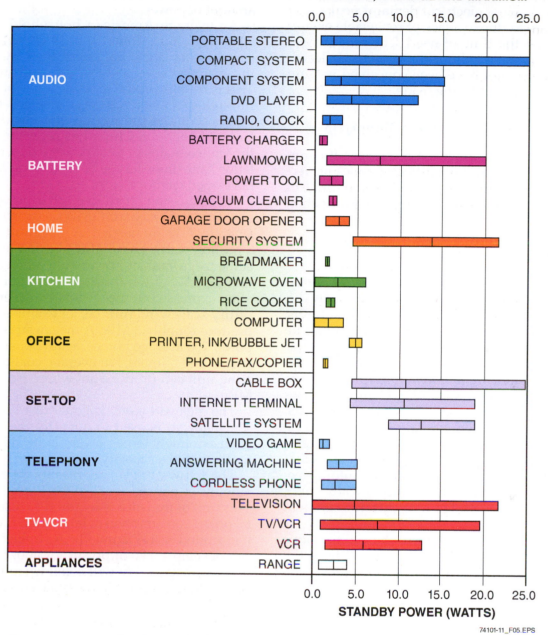

STANDBY POWER (WATTS)
MINIMUM, AVERAGE AND MAXIMUM

Figure 5 Standby power consumption for various devices.

74101-11_F05.EPS

when the cost of fossil-fuel use begins to exceed that of alternative forms. Such an economic situation would surely drive alternative energy use to the forefront—assuming the needed technology is on hand to support it. However, if the cost of fossil fuels were to plummet suddenly, world interest and investment in alternative energy may plummet, too. Cost, therefore, is a strong motivating force independent of the level of technology.

Another obstacle to consider is energy storage. Fossil fuel and nuclear power generation is controllable, allowing power to be produced as needed. Some forms of alternative energy cannot be controlled. The wind blows when it blows, and the sun shines when it shines. Wind and solar power specifically are quite unpredictable. There is no direct relationship between peak electric power demand and peak power generation from most renewable sources. Some means of storing electrical energy is needed. The development of a simple, inexpensive electrical power storage solution would revolutionize the industry. It is not likely to occur in the short term. This makes a comprehensive approach to power generation and control necessary. All the alternatives must be used together in harmony, with fossil fuels, to provide a total solution. Otherwise, power that cannot be used is generated at a significant cost and wasted, while electrical demand peaks may not be satisfied due to a lack of renewable energy reliability at the time of need. The real answer to storing massive volumes of generated electrical power may not have been conceived yet, but progress is certainly being made.

3.0.0 POWER GENERATION ESSENTIALS

As you continue to explore this subject, it is important to understand the different scales of electric power generation. In addition, some frame of reference is needed to quantify power production and use.

3.1.0 Quantifying Power

Although the words *power* and *electricity* are often used interchangeably, they are technically different terms. Electrical energy lights light bulbs, but we pay for power consumed. Power can be described as the electrical energy consumed over some period. The basis for consumption is the watt (W). Incandescent light bulbs typically range from 25 to 150 watts. A handheld hair dryer uses significantly more energy—1,000 to 2,000 watts. An average size liquid crystal display (LCD) TV consumes 100 to 120 watts, while a newer light-emitting diode (LED) TV consumes somewhat less at an equal screen size. Some phantom electrical loads may be measured in milliwatts (mW), which is 1/1,000 of a watt. The kilowatt (kW) is equal to 1,000 watts.

Watts, however, are not purchased as a unit of power. Power is typically purchased from the utility in kilowatt-hours (kWh), which relates the watts of power consumed to the time used. A 100-watt light bulb that is on for one hour consumes 100 watt-hours, or 0.1 kWh, of power. A 50-watt bulb operating for two hours consumes the same amount of power.

An average American household consumes 8,500 to 9,000 kilowatt-hours (kWh) per year, based on statistical data from the EIA. However, the average varies widely by region and climate. This is roughly equal to using 1 kW of electricity through every hour of every day, or 1 kWh. This is almost double the average household use in the United Kingdom. China has recently surpassed the United States in total power use, with a population four times larger but spanning largely undeveloped territory.

As discussions turn to electric power production, the terms *megawatt (MW)*, *gigawatt (GW)*, and *terawatt (TW)* are needed. A megawatt is equal to 1 million watts. Large office buildings may consume megawatts of power annually. The production capacity of wind turbines and solar farms is often expressed in megawatts. An average single utility-scale wind turbine, for example, is rated to produce roughly 2 MW. A megawatt is enough to power 250 average homes, again depending upon region and climate. An average coal-fired generating unit can produce anywhere from 25 to 800 MW.

The gigawatt (GW) is equal to 1 billion watts. The gigawatt is used to quantify the production from extremely large power plants. It is also used to describe the capacity of the power grid.

The terawatt (TW) represents an astounding amount of power, equal to 1 trillion (10^{12}) watts. This unit of measure is required to discuss the total power consumption of humankind. Worldwide, it is around 16 TW annually. A stroke of lightning can develop 1 TW of power for about 3 milliseconds. Perhaps a day will come when we learn to control lightning and harness the energy released.

On Site

Electric Power Storage Research

The DOE's Energy Storage Systems Program (ESS) was created to "develop advanced energy storage technologies and systems, in collaboration with industry, academia, and government institutions that will increase the reliability, performance, and competitiveness of electric generation and transmission in utility-tied and off-grid systems." The program is operated through Sandia National Laboratories.

One solution being studied is the sodium-sulfur (NaS) battery. NaS battery systems have been placed in service for testing as a form of energy storage for renewable wind and solar photovoltaic power. The largest such site is located in Northern Japan, with a storage capacity of 34 MW of electricity and 245 MWh of power available. It is being used to store excess power generated by a wind energy site.

Think About It

How Much Did That Cost?

Examine a recent electric bill for your home. Can you determine how much you pay for one kWh of power? If so, then what is the monthly cost to operate a Microsoft Xbox 360 at 160 watts, together with a 42" LCD TV at 180 watts, for 32 hours each month? Starting hint: Add the watts of the two devices together, and then divide by 1,000 to determine the kW load first.

3.2.0 Generating Facility Scale

When a power-generating project is discussed, the output capacity is often discussed as well. Some idea of scale is needed to fully grasp the discussion. Utility-scale power production (*Figure 6*) is the production of bulk electric power. Power is generated in large volumes for use by anyone connected to the national grid. This large-scale approach is also known as centralized power generation.

The average coal-fired generating unit capacity is roughly 235 MW of power. A centralized power generation plant usually houses multiple units of this size. The volume of electric power production from natural gas is second only to coal in the United States. Simple-cycle natural gas turbines generally range from 100 MW to 400 MW in output. Combined cycle units, which capture and use wasted heat to generate still more power, can produce power in greater volume and at higher effi-

74101-11_F06.EPS

Figure 6 Tampa Electric's Polk Power Station near Lakeland, FL.

ciencies. Nuclear power units average about 1,025 MW of capacity.

Keep in mind that the listed capacity of some generating units may be far higher than actual production levels, except in the case of nuclear power units. Once brought on-line, nuclear generating units remain at 100 percent output. Fossil-fuel facilities have the ability to modulate power generation, and some are much easier to start up or shut down in a reasonable period of time. Some natural gas turbines can be placed in service in a matter of minutes.

Utility-scale power units typically operate as a base load unit or a peaking load unit. Base load units are those that are required to maintain the minimum amount of power needed. Nuclear units, for example, are operated as base load units due to their efficiency and the complexity of starting or shutting down a unit. Base load units and plants operate constantly except when repair or maintenance is required. When electrical consumption rises and additional power is needed, peaking load units are brought on line. Natural gas turbines provide good service for peaking loads since they are much easier to stop and start.

It should be noted that centralized power facilities are not necessarily owned by the utility company that sells power to the consumer. Facilities, especially alternative energy plants, can be privately owned and operated as well. Obviously, there must be a close working relationship between the two groups. This is especially true when alternative energy sources are involved. Utility-scale power generation typically offers the least expensive way to generate power, due to the economy of scale. It also generally results in better emission control than a smaller power plant.

The next level is known as distributed, or decentralized, power generation (*Figure 7*). This approach incorporates smaller power generating facilities that are closer to the electrical load. In some cases, power is generated at the site of its primary use. Distributed generation is often the basis for a microgrid, where several small diverse facilities combine their power to serve nearby loads through a small grid. Distributed power facilities are often alternative energy-based. In most cases, the microgrid is connected to the main grid. It can be disconnected at any time, or draw power from it if necessary. These sites are referred to as customer-sided applications when the power user also generates power.

This scenario provides some good advantages for both parties. If the microgrid generates more power than is needed, it can be sold and transferred to the main grid. The main grid provides a

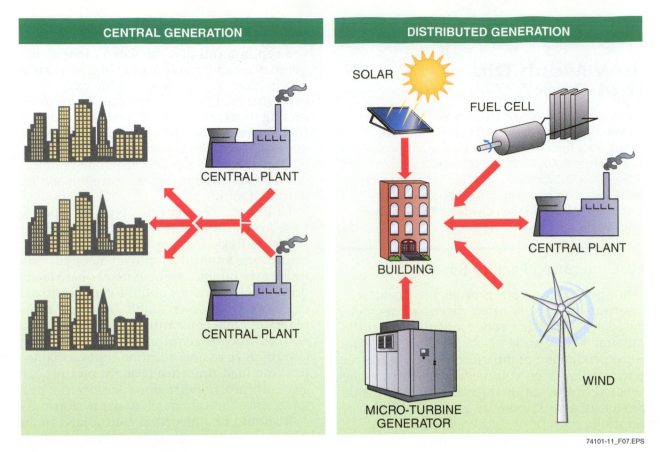

CENTRAL GENERATION	DISTRIBUTED GENERATION

Figure 7 Centralized versus distributed power generation.

back-up power source if there are problems within the microgrid. The microgrid can also be used during peak demand if the local capacity is limited. The distributed generation approach also reduces power losses and the cost of power transmission to the consumer by being close by. The key benefits to distributed generation are the ability to consciously choose the energy source, and back-up sources of power are generally available when needed.

Distributed generation can be a rather costly approach. The benefits outweigh the additional cost in some cases, and some industries have used this energy strategy over the years. The concept is especially attractive to large industrial consumers that have embraced a green mentality and recognize all the benefits it offers, including the admiration of customers of a like mentality.

Alternative energy can be applied in very small applications, such as a single home (*Figure 8*). Solar- and wind-powered systems are readily available for installation. Many utility systems have adopted **net metering** programs and **interconnection agreements**, allowing small systems to be connected to the grid. The concept is simple. When the system is generating more power than necessary, the excess flows into the grid. The electric meter tracks energy flow both into and out

of the grid. At the end of the billing period, the net result of all activity is evaluated and the account is reconciled. Small solar and wind systems can also be fitted with a bank of batteries to store power for use when electric power generation is low or non-existent (*Figure 9*). Small renewable energy systems can be a valuable energy resource for homeowners and businesses.

Figure 8 Residential solar power system.

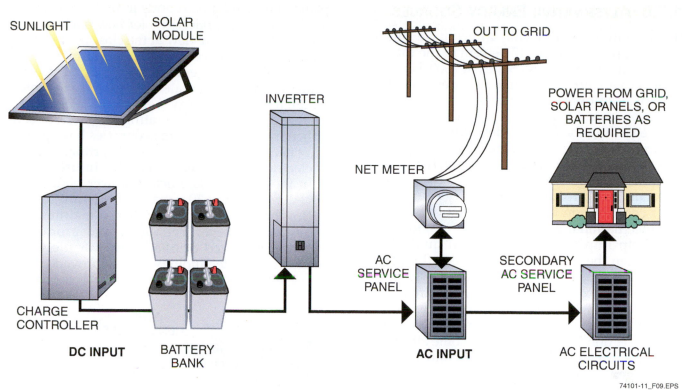

SUNLIGHT **SOLAR MODULE** **OUT TO GRID**

INVERTER

POWER FROM GRID, SOLAR PANELS, OR BATTERIES AS REQUIRED

NET METER

CHARGE CONTROLLER

DC INPUT **BATTERY BANK**

AC SERVICE PANEL

AC INPUT

SECONDARY AC SERVICE PANEL

AC ELECTRICAL CIRCUITS

74101-11_F09.EPS

Figure 9 Grid-connected solar system with battery bank.

3.3.0 Utility Resource Integration

Although utility companies are increasingly required to integrate renewables into their energy portfolio, their consumers expect them to do so as inexpensively and painlessly as possible. Many issues must be considered as the companies make decisions regarding the addition of any source of power. The decision-making process is quite complex and requires serious thought and research. Some of the challenges to be considered include the following:

- *Is there a demand for additional power, and is that demand in a specific area?* Integrating a new resource without new demand means that the utility is technically replacing something that already exists, or creating a redundant resource. Neither situation may be good for customer rates When a specific area needs additional power, the means and distance of transmission from the new source to the new load affect the final choices.
- *What type of resource is in order?* Any power source must be assigned base load or peak load responsibilities. When working with renewable sources, using them for base load service can be very difficult. Renewable sources, such as solar or wind power, operate intermittently at best, but base load demand is consistent. Even though integrating renewables into the energy

mix is desirable, they are not as flexible and consistent as fossil fuel and nuclear-power generating units.

- *When is the resource needed?* The time line for a new power load must be known, and then matched with a resource that can be developed in an acceptable amount of time. Both construction and regulatory processes can be very lengthy and challenging to predict with precision. It is not unusual for regulatory permission for construction to require two years. The construction process can easily require twice that long, and construction cannot begin until permission to do so is granted. Permission to build nuclear facilities often takes far longer, and their construction process is much longer as well. In some cases, whether for economic or practical reasons, utilities must purchase power from other utilities or independent power generators. A wind energy site, for example, may be developed independently and the power sold to the utility at a negotiated cost.

For anyone who believes it is a simple matter to stop using a fossil-fuel power generating unit and switch on a solar installation in its place, the issues noted above should help demonstrate the true complexity of it. These issues only scratch the surface of the difficulties utilities are faced with each day in meeting the demands of both consumers and government entities.

4.0.0 ALTERNATIVE ENERGY SOURCES

It was noted earlier that alternative energy sources provided about 17 percent of the total energy needs in the United States in 2009. However, alternative sources provided 31 percent of the energy used exclusively for electric power generation. *Figure 10* provides information from the EIA regarding the contribution of various energy sources, including the alternatives, for 2010. Note that these figures are related exclusively to electrical power generation.

A continued focus on alternative sources and their contribution to power generation will help advance national energy security. In addition, it will help improve environmental quality and strengthen the energy economy. According to the EIA, renewable energy is the only power-generating source expected to show significant growth by 2035. The percentage of all other sources is expected to either shrink or remain relatively stable.

This section briefly introduces some alternative energy sources and their present contribution to power generation. Their potential is also examined.

4.1.0 Nuclear Power

The DOE's Office of Nuclear Energy (NE) promotes the use of nuclear power as a key source of clean power. Nuclear power provides roughly 20 percent of the United States' present electrical power. Small reactors also power ships and submarines, but present technology and safety concerns do not yet allow for their widespread use in transportation. Beyond the production of electric power, nuclear energy is also used for medical treatments and diagnostic systems.

Although nuclear power provides 20 percent of electrical power, it provides 70 percent of the power that does not produce carbon emissions. Although there are safety and security concerns related to nuclear materials, nuclear power produces no greenhouse gases. It is a source that is controllable, continuous, and capable of producing large quantities of electricity. There are currently 104 operable nuclear power generating units in the United States. Over the years, 28 have been shut down or retired, according to statistics from the EIA. Only two new reactors are currently under construction in the United States, located in eastern Tennessee and Georgia. The construction of Unit 1 at the Watts Bar facility was started in 1973, and the unit was brought into service in 1996. A second unit was planned, but those plans were scrapped in 1988. In 2007, the decision was made to restart construction of Unit 2, and its startup is presently

planned for 2012. EIA reports indicate there are 24 active licensing applications for new nuclear units that remain under review at the close of 2010. Four other applications were recently withdrawn.

Nuclear power generation (*Figure 11*) is also growing on a global scale. According to the International Atomic Energy Agency (IAEA), in 2011, there are 65 nuclear-generating units under construction worldwide. Some countries have placed their faith in nuclear power as the primary means of electric power generation for the future.

In the early 1990s, nuclear power plants produced power only 70 percent of the time, due to both scheduled and unscheduled maintenance activities. However, reliability and productivity have improved dramatically over the years, thanks to new technology and improved procedures. A unit in Illinois presently holds the record

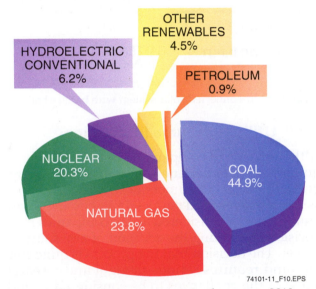

74101-11_F10.EPS

Figure 10 US electric power generation by source, 2010.

74101-11_F11.EPS

Figure 11 Cooling towers of a nuclear power station.

for uninterrupted operation of 739 days, as documented by the Nuclear Energy Institute (NEI), when scheduled refueling finally halted its operation. As of 2010, NEI and the EIA both reported that the nuclear unit capacity factor, which compares total nuclear power capacity with its actual output, had risen to 90.5 percent.

There is no doubt that issues like security and waste disposal present challenges to its wider use. Opponents to nuclear power focus on these issues, as well as safety concerns, for the future. However, it is evident that nuclear power is here to stay and is a necessary part of an energy portfolio. It already makes a significant contribution to electrical power supplies that cannot be replaced by another source at this time. The case to expand nuclear energy use is compelling. Its overall contribution to electrical power generation is expected to decline only 1 to 2 percent by 2035. This is primarily due to the projected growth of renewable sources, rather than any significant reduction of nuclear power capacity.

4.2.0 Hydroelectric Power

Of the sources considered renewable, water contributes the most to power production presently. Hydroelectric power is generated by the forces of flowing or falling water to turn a turbine. The typical method is to dam a stream or river, taking advantage of the weight of the water stored behind the dam. Water is guided down through the dam and spins a turbine, and then continues downstream. The turbines turn generators to create electrical power.

Hydroelectric plants have long economic and mechanical lives. They are also simple to operate and require little or no fossil fuel to maintain. Since fossil fuels are not involved, there are no carbon or greenhouse emissions resulting from their operation. Of course, Mother Nature primarily determines where a major facility can be placed at a reasonable cost for the power produced. Topography and elevation make hydroelectric power possible, while additional factors affect its practicality. New technology may allow the use of new sites that have been previously considered impractical options. Hydroelectric plants often have a major effect on the local environment. The effects can be both positive and negative. Some of the best recreational destinations in the United States have been created by hydroelectric dams.

The world's largest hydroelectric power capacity is presently found at Three Gorges Dam in China, which will generate 22,500 MW of power when the generators are fully operating. The largest facility in the United States is the Grand Cou-

lee Dam (*Figure 12*), originally completed in 1942. It is located on the Columbia River in Washington. This facility is ranked fifth in the world at 6,809 MW. The United States has no other facilities ranked in the top 25 except the Chief Joseph Dam, ranked 25th. It was also constructed on the Columbia River in Washington.

Hydroelectric plants provide about 7 percent of the United States' total electricity needs now. In the early part of the twentieth century, hydroelectric plants accounted for nearly half of the United States' electric power. Utility-scale expansion in hydroelectric power is not expected. The simple fact is that most of the best locations for large facilities have already been developed. By 2035, hydroelectric power will likely provide an even smaller portion. The most likely growth will be in smaller facilities that operate as part of distributed microgrids. There are certainly advantages to clean hydroelectric power on any scale.

4.3.0 Wind Power

People living in areas with poor wind resources know very little about wind power. On the other hand, those who live in the shadow of the giant turbines are very aware of them. Although the southeast region of the United States has relatively poor wind resources, some parts of the country have an excellent resource. Wind provides utility-scale power from sites known as wind farms or wind energy sites (*Figure 13*). Wind can also be used to generate power for remote, off-grid locations, and sites that are grid-interactive. Power generated by wind is emission free.

Wind power presently accounts for about 2 percent of US electrical power today. About 35,000 MW of power is now being generated by wind turbines. The DOE has set a goal of providing 20 percent of power from wind by 2030. There are some indications that the goal is realistic, but

74101-11_F12.EPS

Figure 12 The Grand Coulee Dam and bridge.

Figure 13 Wind energy site.

74101-11_F13.EPS

many factors will affect the result. Many planned wind facilities have been unable to develop. The reasons vary, but local resistance and environmental issues are significant obstacles.

As of 2009, the United States was generating more electrical power from wind than any other country. Germany and China were second and third, respectively. Due to the distribution of wind resources, some US states are generating a great deal more power than others. At the end of 2010, Texas easily led the way at over 10,000 MW. Texas was also the site of the world's largest wind farm at that time. With 627 turbines, the Roscoe Wind Farm can generate 781.5 MW. Iowa was second by generating 3,675 MW of wind power. A report released by the NREL in 2010 revealed that US onshore wind resources had the potential to generate more than 10 million MW of power. This would result in power generation nine times larger than the current needs. Local opposition to wind farms will likely prevent a great deal of these resources from being used.

Great wind resources are available offshore too. However, the installation cost is significantly higher. Construction of the first offshore wind energy site serving the United States was scheduled to begin in 2011 near Nantucket, MA. There is no doubt, though, that a great deal of effort will continue to be placed in wind energy in the future. The resource is here, but how much of it can be used remains to be seen.

4.4.0 Solar Power

Solar energy is another emission-free power source. The sun also provides light and heat that can be used directly. Solar **photovoltaic (PV)** systems (*Figure 14*) use sunlight to produce electricity without emissions. Solar PV systems can be utility scale or small enough to power a landscaping light. Utility-scale systems are well behind wind power, but they are advancing. The cost of photovoltaic power generation from solar energy has fallen dramatically over the years. The primary cost is the solar panel.

GOING GREEN

Arizona's Agua Caliente Solar Project

In January 2011, the DOE announced its support through a conditional loan guarantee of nearly $1 billion for a solar project in Yuma County, Arizona. The PV project is expected to generate 290 MW of solar-generated power in Phase 1, providing enough green power for 100,000 homes at its peak operating capacity. In the process, it is expected to eliminate 237,000 metric tons of greenhouse gas emissions per year. This is comparable to removing 40,000 vehicles off the road annually. At the time of the announcement, the project would be the largest solar project in the world.

Figure 14 Solar PV array.

74101-11_F16.EPS

Figure 16 Concentrated solar thermal (CST) test facility.

Most of the discussion of solar energy is focused on electric power generation. However, the sun's energy can also be used to heat water or create steam. These systems are referred to as **solar thermal energy (STE)** systems (*Figure 15*). Water or another solution is pumped through solar collectors exposed to the sun. Low- and medium-temperature systems heat water for swimming pools or other domestic uses. Research continues on utility- and distributed-scale systems that create steam to generate electrical power. Known as **concentrated solar thermal (CST)** systems, they use a system of mirrors to focus the sun's rays on a collector (*Figure 16*). These systems use this focused energy to create and store steam, and the steam is then used to drive a turbine. These systems are ideal for installation in desert areas where land and sunshine is plentiful. Producing and storing the steam during peak solar hours allows power to be generated even after nightfall.

So far, utility-scale solar PV systems contribute very little power to the grid. CST system concepts and capacities are better suited to utility-scale power generation than PV systems. Their cost to construct and maintain is presently quite high. California's Mojave Desert is home to one of the world's largest solar power facilities (*Figure 17*), with a generating capacity of 354 MW. Its average production so far is just 21 percent of that amount.

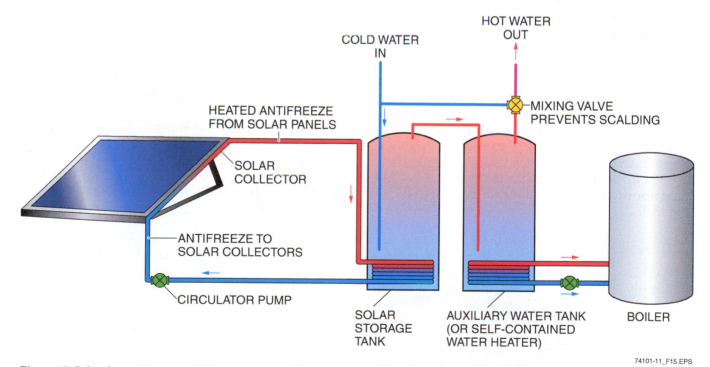

Figure 15 Solar thermal energy (STE) system example.

Figure 17 Solar energy generating system (SEGS) facility.

The facility covers 1,600 acres with nearly one million mirrors. Around 3,000 mirrors per year must be replaced, mostly due to wind damage. Power generation from solar energy overall is still relatively expensive when compared to other forms. According to estimates from the DOE's Energy Information Administration (EIA), power from solar energy will still cost 3 to 5 times more per megawatt than fossil fuels by 2016.

However, many small solar systems are in operation across the United States. Many are connected to the grid through net metering programs. One way the DOE supports solar power is through the Solar America Initiative (SAI). Launched in 2007, the goal is to make solar power competitive in cost with current sources. Another program, the Solar America Cities program, is a partnership with 25 US cities to develop an organized approach to solar energy use on a distributed scale.

4.5.0 Biomass

Biomass is any material produced from living or recently living things and their byproducts. Wood, for example, is a large source of biomass energy. Simply building a fire for warmth is an example of converting biomass to heat energy. Biomass, primarily wood, is often co-fired with coal to generate power. At present, co-firing biomass with coal is considered the most efficient use of biomass for power generation.

The heat from biomass incineration is used in a number of different ways. It can produce steam to drive power-generating turbines, or provide heat for industrial processes. The majority of research in this area is devoted to increasing functional efficiency and reducing harmful emissions at an acceptable cost.

Although not all waste is classified as biomass, the burning of solid waste is generally considered the use of biomass energy. Beyond the fuel source, waste-to-energy units operate much like coal-fired generating units.

About 1 ton of waste must be burned to equal the heat output of 500 pounds of coal. The biomass-to-energy process does cost more than coal. However, one big advantage comes from the reduction of solid waste. Every ton of solid waste burned for energy is a ton that does not need to be buried.

Through a process known as gasification, wide varieties of biomass types are burned at extreme temperatures to produce a fuel gas. The fuel is then used to power gas turbines and generate power. The DOE has sponsored a number of different gasification systems and studies across the country with utility companies, each of which is a candidate for commercialization as a significant power-generating unit.

The Solar Decathlon

GOING GREEN

74101-11_SA02.EPS

The DOE-sponsored Solar Decathlon invites collegiate teams to design, construct, and operate solar-powered homes. The homes must be cost effective, energy efficient, and attractive. The winning team is the one that creates the best combination of affordability, appeal, energy efficiency, and power production.

International interest in the event, which began in 2002, has been very high. The 2011 competition included teams from Belgium, Canada, China, and New Zealand.

Landfills can produce a fuel gas through the anaerobic digestion of biomass. The term *anaerobic* means that air is absent. Biomass and refuse buried in a landfill lack a source of air, and a bacterial-fermentation process begins that results in a fuel gas. The collected gas is typically 40 to 75 percent methane. After some purifying and processing, the gas can be used to power generators, and it can even be fed into the national natural gas supply system. Anaerobic digestion can be duplicated outside of the landfill environment to generate power or provide heat on a smaller scale.

Biomass is also a popular fuel for cogeneration. Cogeneration can be defined as the use of energy to provide both heat and electrical power. A sawmill, for example, may use the sawdust, bark, and shavings it normally produces as waste to create heat and generate electrical power for on-site use.

Biofuel is a fuel that begins as biomass. Two examples of biofuels are ethanol, made from corn, and biodiesel. Biofuels are growing in popularity to power emergency generators as a source of backup power for critical facilities. This is historically done with petroleum-based diesel fuel, but biofuels provide a significant reduction in emissions.

Biofuels are getting plenty of attention along with biomass. Replacing fossil fuels in motor vehicles is a big part of the US energy plan. The Renewable Fuel Standard enacted in 2007 requires that biofuels supply at least 21 billion gallons of US motor fuel by 2022. Biofuels like ethanol have been around for some time now and are used heavily. They are typically blended with gasoline. Nearly all gasoline today is 10 percent ethanol, and higher percentage blends will soon be in use.

Biofuel sources are a specific subject of research. Corn has been used for many years now. Many other sources, such as switchgrass and algae, are being tested. Fast-growing poplar trees are being grown specifically for biomass energy production (*Figure 18*). As time goes on, using an important food source such as corn for this purpose may not be the best idea. A non-food source material that is cheap and fast growing may be better.

About 53 percent of biomass energy is used by various industries. Another 22 percent of it goes to transportation fuels. It is currently used by utilities to provide only about 1.5 percent of US electricity. However, new technology to reduce

Waste Generation

Each American generates roughly 4.6 pounds of solid waste each day. Over 33 percent of that was recycled, and 54 percent was dumped in a landfill in 2008. The rest was either composted or incinerated. It may not be perfect, but consider the figures from 1980. At that time, we recycled less than 10 percent, while 89 percent of it went to landfills. More and more waste is finding its way into new products or becoming a source of energy. *Reduce, reuse, and recycle.*

GOING GREEN

74101-11_SA03.EPS

emissions and improve combustion may help biomass become more popular as time goes on. Solutions are needed for waste disposal as well as new energy sources.

4.6.0 Other Alternative Sources

There are many other alternative ways to create or capture power and heat for use. New ideas are tested daily, while others are already well known. Most represent opportunities for power generation on a small or distributed scale. Here are some examples:

- *Geothermal energy* – Geothermal energy comes from the heat of the earth itself. The extreme temperatures found deep beneath the earth's crust cannot be easily reached. However, what can be reached is used to generate electricity and provide hot water (*Figure 19*). It is classified as renewable since rain returns any water removed. The slow decay of radioactive particles helps to maintain the heat. About 0.4 percent

of consumed energy comes from geothermal sources. Geothermal heat pump systems are used in homes and commercial buildings for heating, cooling, and hot water production. They take advantage of the stable temperatures found just below the surface for heat transfer.

- *Fuel cells* – **Fuel cells** can generate electricity from a chemical reaction. Although there are many types, the hydrogen fuel cell is the most promising at this time. Hydrogen is used as the fuel, and oxygen supports the reaction. Fuel cells can be both compact and lightweight. A lot of research continues into their use as a source of electricity to power vehicles (*Figure 20*). Stationary fuel cells have been deployed successfully at hospitals, nursing facilities, and other critical locations as a back-up power supply. One interesting application is a subway maintenance shop in Queens, NY. A 200 kW fuel cell there powers maintenance and cleaning equipment, and powers lighting systems during grid power outages. The residual heat from the fuel cell process is captured and used to heat domestic hot water in the same facility. This is an example of a CHP process, where energy is used to both generate electricity and process heat.

- *Wave and tidal energy* – Wave and tidal energy is the electrical power created from wave and tidal action. Wave systems (*Figure 21*) can be placed on top of the water or deeper to capture the continuous motion of water. The total potential of wave power is estimated to be about 7 percent of current electrical needs. The efficient capture and transmission of this energy is the point of research. Tidal power relies on the in-and-out cycle of the tides. As shown in *Figure 22*, a rotor like that of a wind turbine turns as water surges towards land. As the tide recedes, the rotor turns in the opposite direction.

74101-11_F18.EPS

Figure 18 Poplar trees grown and harvested for their biomass.

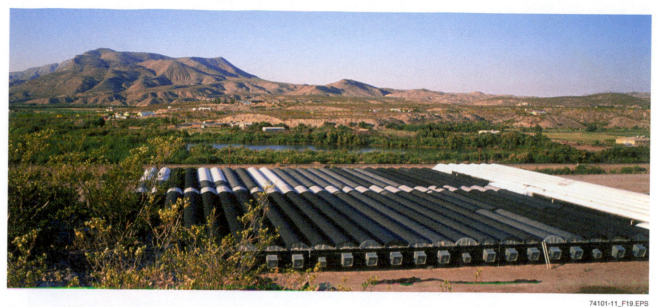

Figure 19 Greenhouses in New Mexico heated by geothermal energy.

74101-11_F19.EPS

5.0.0 THE ELECTRICAL GRID

The system of electrical power transmission and distribution lines across the United States is known as the grid (*Figure 23*). In the last decade, the word has become more familiar to Americans. There are concerns regarding the age and overall design of the grid as attempts are made to integrate new distributed power sources. However, it is quite functional in spite of its age.

Technically, the US power grid is the largest joined machine on Earth. Some have even referred to it as an ecosystem due to its complexity. There are well over 9,000 electrical generation units (not including small net metering systems) and one million MW of capacity. The transmission lines extend 300,000 miles. The National Academy of Engineering recognizes the grid as the most significant achievement of the twentieth century.

During the first 10 years of the twenty-first century, grid transmission lines have only been extended 668 miles. The DOE estimates that power outages and related issues cost Americans $150 billion every year. Imagine what the cost would be if it worked at less than its 99.97 percent level of reliability. This level of reliability in a system of its size is quite remarkable.

Remember that power must be used as it is generated. There is no means of mass storage at this time. Existing fossil-fuel power plants are controllable, producing power as the need arises. So far, the grid is keeping up with the generation of power. However, new generation methods and power sources are challenging to integrate. In many cases, this is due to the unreliable production of power from renewable sources, rather than a weakness in the grid itself.

5.1.0 Grid Risks and Concerns

Some specific issues regarding the existing grid are efficiency, economics and reliability, security, and alternative power generation.

5.1.1 Efficiency

There are losses associated with operation of the existing grid. One DOE study focused on the result of raising the transmission efficiency of the grid just 5 percent. It was found that the energy saved would be equal to eliminating the fuel and emissions from 53 million cars. Consider the gains from a grid that is 25 percent more efficient. There is much to gain from improving grid efficiency. Part of the solution is to shorten transmission distances. Distributed power generation has this very effect. As time goes on, new technology

74101-11_F20.EPS

Figure 20 NREL hydrogen fuel cell demonstration vehicle.

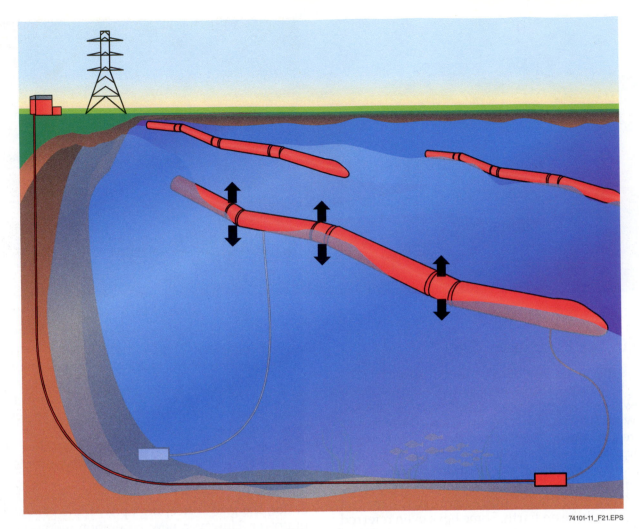

Figure 21 The Pelamis wave power device off the coast of Portugal.

74101-11_F21.EPS

may help identify ways to increase the efficiency of bulk power transmission.

5.1.2 Economics and Reliability

The reliability of the grid and economic losses are related. Some of the losses related to grid failures can be staggering:

- In 2000, a one-hour power outage affected the Chicago Board of Trade. The resulting delay held up $20 trillion in trades.
- Sun Microsystems says that a power outage costs the company $1 million each minute.
- A blackout that occurred in the northeastern United States in 2003 resulted in an economic loss of roughly $6 billion.

Beyond the financial losses, grid reliability affects many common functions of daily life. Grid problems can cause massive spoilage of frozen foods, traffic light system failures, and the halt of personal financial transactions. Life without

power is uncomfortable and potentially dangerous. In spite of the losses that can occur because of power failures, a reliability level of 99.97 percent will be very difficult to exceed.

Improving the grid has the potential to affect the US economy in positive ways. Bringing green alternative energy to the grid will require a lot of work by people with specific skills.

5.1.3 Security

The security of the power grid is essential. Taking a few key sections of the grid offline at the same time would have a dramatic effect. The failure of one section could cascade to other sections very quickly. Even communications and banking systems that remain powered would become nearly useless due to the loss of connectivity. However, securing the largest joined machine on Earth is easier said than done.

The North American Electric Reliability Corporation (NERC) has the mission of ensuring the re-

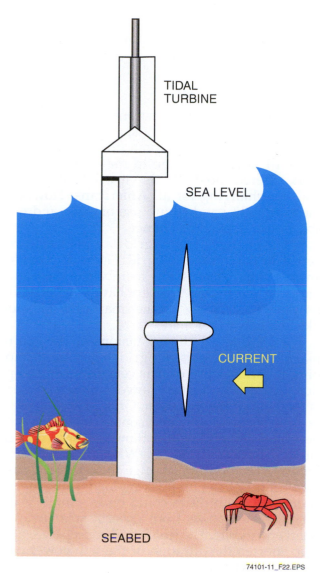

Figure 22 Tidal turbine.

Figure 23 Grid substation.

liability of the North American power system as a whole. Its Critical Infrastructure Protection program coordinates all of the organization's efforts to improve physical and cyber security of the bulk power system relative to its reliability. It provides assessments of risk and preparedness, and shares critical information through an alert system with industry. Although not a government organization, NERC has far-reaching power and is subject to the oversight of the Federal Energy Regulatory Commission (FERC), as well as Canadian entities. In recent years, there have been reports of attempted hacking into utility control systems. The coordinated efforts of NERC have enabled such threats to be analyzed and made harmless quickly.

As utility companies become more dependent on the Internet to communicate and function, new avenues for access become available. Since the Internet will continue to play a large part in

power control, great care must be taken to secure networks.

5.1.4 *Alternative Power Generation*

The grid has been designed and built using power from large, centralized generation sites. Current fossil-fuel plants are quite controllable, allowing power to be generated as needed. Nuclear units provide reliable and predictable power for base loads, while fossil-fuel units handle both base and peaking load duty just fine. These are reasons the grid has functioned so well to date—reliable and predictable power generation, coupled with a well-conceived plan of use.

Power from some renewable sources is not so controllable or reliable. While a fossil-fuel plant can increase or decrease output quickly, wind energy and solar PV sites cannot. In addition, power lines must be brought to the many new locations of renewable power generation to make use of the power. That investment alone is substantial for a power source of lesser reliability. The grid needs to be modernized somewhat to allow alternative sources to be used to their maximum potential. Information and data collection systems needed to make wise and quick decisions in the use of

---

Think About It

Reliability

The US electrical grid is a source of concern, even though it operates at a 99.97 percent level of reliability. How many other systems or objects can you think of that operate at such a high level of reliability? Would there be a problem if they were 0.03 percent less than perfect?

---

many new power resources is not presently available. The present grid design does not allow for the most effective use of distributed power generated by renewable sources. The grid should not necessarily be considered deficient; it simply was not designed to accept power generated from small, scattered locations.

On Site

US Northeast 2003 Blackout

These two photos from NASA's Earth Observatory show the difference in lighting before and after the blackout. The investigation revealed that power flowed normally north of the Great Lakes, from east to west in the late afternoon. Then power flow suddenly reversed its direction and doubled in volume. Soon, computer programs in 21 power stations detected the cascading problem and shut down. This was automated to protect the grid from the overload. It was later determined that several minor problems led to one big one under the strain of summer heat and excessive power flow.

August 14, 2003, 9:29 PM EDT, About 20 hours before blackout

August 15, 2003, 9:14 PM EDT, About 7 hours after blackout

74101-11_SA04.EPS

5.2.0 The Smart Grid

The weaknesses and limitations of the grid must be addressed. What is envisioned is something much better. Attitudes about power and energy use will have to change to make the solution work. There is no doubt that it will be expensive and life-changing for all Americans. However, the availability and future use of clean and reliable power depend on change.

Major changes to the grid are packaged together in the term **smart grid**. *Figure 24* provides an idea of the concept. Key members of the power industry and the government have been meeting for years to discuss what the smart grid will look like and how to get there.

The DOE and industry visions of the future make use of the existing grid structure. Scrapping it and starting over is certainly not an option. A fully automated power delivery system is the goal of the program, Grid 2030. A system capable of monitoring and controlling every connection to the grid would allow power to flow in either direction seamlessly. When power loading increases, rather than change the output of base load units or start up peaking load units, the response by the utilities could be as simple as diverting excess power from a nearby solar array or wind energy site. The smart grid concept will also help shorten the duration of the few power outages that do occur.

Note the small antennas in *Figure 24* positioned near electrical sources and loads. Under the smart grid concept, the consumer becomes a part of the system. Homes and businesses would be connected to the utility by two-way digital communication. Smart appliances can be programmed to operate under priorities established by the owner. Home-energy monitoring displays will provide consumers with real-time information. Using power during off-peak hours of consumption will be billed at a reduced rate. In this way, consumers can decide how and when to use power to fit their lives.

The smart grid is based on two-way communication between electrical loads and power sources. Remember that power must be consumed as it is generated. The smart grid will provide for coordinated control of all power sources. In addition, the system could choose which resource to use based on cost. Loads, such as appliances or non-essential plant equipment, would also be controlled by the utility with permission. Some measure of

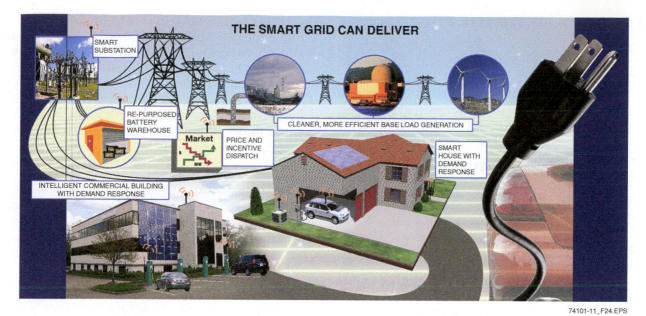

THE SMART GRID CAN DELIVER

SMART SUBSTATION

RE-PURPOSED BATTERY WAREHOUSE

Market

PRICE AND INCENTIVE DISPATCH

CLEANER, MORE EFFICIENT BASE LOAD GENERATION

SMART HOUSE WITH DEMAND RESPONSE

INTELLIGENT COMMERCIAL BUILDING WITH DEMAND RESPONSE

74101-11_F24.EPS

Figure 24 Smart grid concept.

control over loads, along with full control of the resources, makes an intelligent system indeed. When consumer flexibility and choice is added to the plan, everyone benefits.

5.3.0 Alternative Energy Integration

One of the primary goals of the smart grid is to integrate alternative energy sources with mainstream power distribution. Many areas are already embracing a concept known as the zero-energy district. The concept combines energy efficient, environmentally friendly design with local renewable power to achieve zero use of grid power. There are many different versions of the idea already in practice. However, the basic concept remains: reduce or eliminate carbon emissions from fossil-fuel sources and maximize efficiency and green construction methods.

Grid power may be used by zero energy locations at some point during peak demand. However, at all other times, the provision of clean power to the grid should offset that short-term use. The smart grid can take advantage of the concept by controlling the power source. When a zero energy facility needs grid power, excess power generated by a nearby zero energy facility can potentially fill the void. Zero energy districts can work together to share power as needed.

Zero energy districts and facilities may be one of the best ideas to integrate alternative energy sources. Decentralizing power sources and controlling them all wisely through a smart grid will reduce transmission losses and allow loads and sources to be matched with higher precision.

6.0.0 ALTERNATIVE ENERGY CAREER OPPORTUNITIES

Career opportunities in alternative energy include most of the careers related to the power industry. Many of the same skills and characteristics are necessary. However, each source of alternative energy does require some specialized skill sets. Career opportunities exist at every education level. It is important to note that alternative energy has already brought new jobs and revenue to rural areas that offered very few jobs for residents in the past.

According to the Center for Energy Workforce Development (CEWD), a number of workforce positions in the energy industry are facing shortages due to the retirement of existing workers. The organization has developed competency models to identify the areas of competency required for power generation, transmission, and distribution workers. This information helps to identify candidates who already possess the necessary skills for energy industry employment, and to guide those who do not possess the skills towards competency in critical areas. The power industry offers very stable careers, and jobs that cannot be exported.

The following sections identify and present a few of the career fields that exist in the energy industry.

6.1.0 Power Line Workers

Regardless of the energy source, power must be transmitted over long distances and distributed to the consumer safely and reliably. Each alternative energy source may require slightly different electrical skills and offer different experiences. Electrical power line workers are shown in *Figure* 25 installing a solar module for long-term testing. Line workers install and repair the cables and other crucial transmission and distribution components. Power line workers are typically separated by the portions of the system they service and construct. The classifications include the transmission system, with voltages generally equal to or greater than 115 kV; distribution sys-

GOING GREEN

Smart Grid City

Cities across the country are trying out the new technology in a variety of ways. Xcel Energy has created Smart Grid City in Boulder, CO and is working with what is possibly the greatest number of smart grid features to date. It includes a digital, high-speed broadband communication system; upgraded substations, feeders, and transformers; smart meters; and Web-based tools available for consumers. Find out more about the test program at http://smartgridcity.xcelenergy.com/.

Xcel Energy

The Smart House

Xcel Energy's Smart Grid Consortium is imagining a future that would allow you to communicate your energy choices to the power grid and automatically receive electricity based on your personal needs.

The potential benefits:
- Lower cost of power
- Cleaner power
- A more efficient and resilient grid
- Improved system reliability
- Increased conversation and energy efficiency

Plug-in Hybrid Electric Car
Xcel Energy is studying how plug-in electric vehicles can store energy, act as backup generators for homes and supplement the grid during peak hours.

Smart Meter
Real-time pricing signals create increased options for consumers.

Smart Appliances
Smart appliances contain on-board intelligence that "talks" to the grid, senses grid conditions and automatically turns devices on and off as needed.

High-Speed Connections
Advanced sensors distributed throughout the grid and a high-speed communications network tie the entire system together.

Customer Choice
Customers may be offered an opportunity to choose the type and amount of energy they'd like to receive with just the click of a mouse on their computer.

100 percent green power? A mix of sources? The cheapest priced source? In Smart Grid City, it could be up to you.

Smart Thermostat
Customers can opt to use a smart thermostat, which can communicate with the grid and adjust device settings to help optimize load management. Other "smart devices" could control your air conditioner or pool pump.

74101-11_SA05.EPS

GOING GREEN

FortZED

Fort Collins, Colorado has started an effort to build a zero energy district. The plan includes the downtown area and the main campus of Colorado State University. More than 7,000 residential and commercial power consumers are part of the project. The new solar array at New Belgium Brewing, one of the program participants, is only one of their many renewable energy efforts. Find out more about FortZED at www.fortzed.com.

74101-11_SA06.EPS

tem workers, servicing the system that distributes power to end users; and substation workers, who focus on these critical junctures in the grid.

In the future, smart grid technology and an expanding grid will increase the need for well-educated and skilled power line workers. A high school diploma or equivalent is the minimum requirement for entry-level workers. A two-year degree or technical certificate will allow for more rapid advancement and entry into the profession at a higher level.

6.2.0 Technicians

Wide varieties of technical positions are available in the power and alternative energy industries. The list includes power plant operators, electricians, instrumentation and control technicians, power maintenance and repair technicians, relay technicians, meter technicians, and combustion gas service specialists. Opportunities also exist

Figure 25 Power line worker.

74101-11_F25.EPS

in the growing fields of solar PV, solar thermal, and wind turbine system installation and maintenance. Such jobs provide great opportunities for people who enjoy working with both their hands and their minds.

Energy auditors are often considered technicians as well, since they inspect and document residential, commercial, and industrial energy operations on site. An energy audit reviews all the potential points of building energy loss, such as pathways that allow air to move freely in and out (*Figure 26*). Once an inspection is complete, an energy auditor can then make recommendations to resolve losses and save energy. Once work is completed by the client or others, auditors may inspect the work and ensure that it meets accepted standards of quality and functionality.

As the smart grid concept grows, utilities will need additional administrative workers with technical backgrounds. Power customers will require additional support and information to make wise decisions in the control of their power usage. Technical assistance for customers who own and operate grid-interactive solar or wind systems, and those who wish to consider their installation, is also required.

A high school diploma or equivalent is required for entry-level positions. However, a two-year degree or technical certificate will allow for more rapid advancement and entry into the profession at a higher level. Some positions may require a degree at the entry level.

6.3.0 Construction Trades

Power generating facilities, regardless of the fuel source, must be built by someone. The diverse forms of alternative energy make each type of facility different. Skilled craft workers (*Figure 27*) are needed to construct and maintain such facilities. In many cases, craft workers will need additional training and exposure to new methods and

---

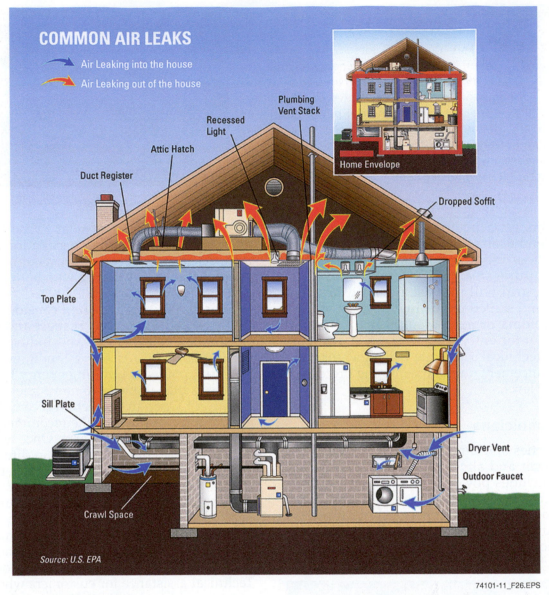

COMMON AIR LEAKS

→ Air Leaking into the house
→ Air Leaking out of the house

Plumbing Vent Stack

Recessed Light

Attic Hatch

Duct Register

Dropped Soffit

Top Plate

Sill Plate

Home Envelope

Dryer Vent

Outdoor Faucet

Crawl Space

Source: U.S. EPA

74101-11_F26.EPS

Figure 26 Home air movement pathways.

materials to construct highly advanced power facilities, wind turbines, and solar arrays. Boilermakers and pipefitters are both crafts that are expected to be in demand, in addition to electricians and wind energy technicians.

A high school diploma or equivalent is the minimum requirement for entry-level workers. A two-year degree or technical certificate will allow for more rapid advancement and entry into the profession at a higher level.

6.4.0 Engineers

Engineers have some of the highest starting salaries of all college graduates. In basic terms, they use science and mathematics to develop solutions to technical problems. Their work connects scientific discoveries with real applications that meet consumer needs. Engineers are needed in the following disciplines to support the use of alternative energy and to improve the fossil fuel technology of today:

- Chemical
- Civil
- Computer
- Electrical
- Electronic
- Environmental
- Geotechnical
- Health and safety
- Hybrid, such as computer sciences/electrical, and telecommunication
- Material
- Mechanical
- Nuclear
- Power

Figure 27 NREL research facility construction.

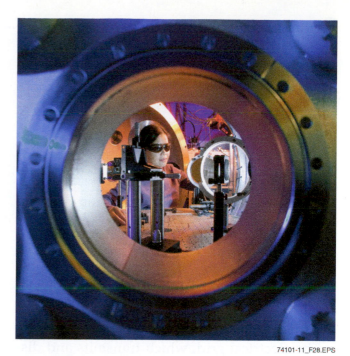
Figure 28 Biomass chemist.

Engineers from all these disciplines are an essential ingredient of the entire energy industry. They are expected to create better ways to generate, transmit, and distribute electric power. While science creates the foundation for nuclear, wind, solar, and biomass-generated power, engineers bring science to life in its practical application. A bachelor's degree is required for entry into most of these careers.

6.5.0 Chemists and Material Scientists

Chemists and materials scientists are needed for research and development to bring new technology into view. Chemists (*Figure 28*) are always working to improve use of fossil fuels to produce cleaner energy. Their work in alternative energy must include a variety of possible sources as they investigate how chemistry can take better advantage of them. Material scientists study the use and practical application of new **superconductors**, fuel cells, and integrated circuits. A bachelor's degree is required for all scientific disciplines. Advanced work may require a master's or doctorate degree.

On Site

Economic Potential

A 2010 study by the Washington Economics Group focused on Florida's possible development of 700 MW of renewable power. The study found that it would likely generate 40,000 new jobs over time. In addition, economic activity in the state would expand by $8.1 billion.

SUMMARY

Alternative energy usage is an essential part of the future. Fossil fuels continue to serve, but energy security and a desire to eliminate harmful emissions demand that alternative sources be developed.

The efficient use of power is another strategy that must be explored. There are many obstacles to alternative energy use ahead, making energy conservation even more important. Regardless of the state of energy sources, energy conservation and efficiency is a wise pursuit.

A variety of alternative energy sources are available. None of them alone can support the growing need for power. A mix of resources, including fossil fuels, are required for many years to come. One of the greatest challenges ahead will be to develop ways to use all energy resources together in harmony.

The electrical grid, which transmits and distributes electrical power, is an important factor in the greater use of alternative energy. In order to make use of alternative sources, the power they produce must be transported to the grid. Since the grid was originally developed around the concept of large, centralized power generation, a great deal of planning and investment is required to make use of alternative power generated at small, widely distributed locations. Technology has not yet made the storage of massive amounts of power possible, further complicating the integration of renewable energy sources that are less reliable and predictable. The smart grid is one way to use power more efficiently, and help integrate electric power from alternative sources into the plan.

The world of energy and electrical power has always offered many diverse opportunities for careers and employment. With new sources and power-generating concepts being consistently added, the opportunities continue to expand. New careers in science, engineering, administration, and the craft industries are being identified along the way. The possibilities provided by the power industry have never been more diverse and unique.

1. Assuming all conditions remain exactly the same, about how many years of known global oil reserves remain at present rate of use?

   a. 15
   b. 25
   c. 40
   d. 60

2. All alternative energy sources are also considered renewable.

   a. True
   b. False

3. How much electrical power does an average American household consume every hour?

   a. 1 kW
   b. 1 MW
   c. 100 watts
   d. 8,500 watts

4. A gigawatt is equal to 1 billion watts.

   a. True
   b. False

5. An average nuclear power generating unit has a capacity of _____.

   a. 84 MW
   b. 235 MW
   c. 660 MW
   d. 1,025 MW

6. How much of US electric power generated in 2010 came from alternative sources?

   a. 8 percent
   b. 12 percent
   c. 17 percent
   d. 31 percent

7. Of the electrical power that is produced without carbon emissions, nuclear power contributes _____.

   a. 20 percent
   b. 45 percent
   c. 70 percent
   d. 85 percent

8. The primary reason for the limited future expansion of hydroelectric power in the United States is because _____.

   a. the facilities cost too much to operate and maintain
   b. the environmental damage is considered too great
   c. most of the ideal locations have already been developed
   d. it does not provide a significant amount of power per facility

9. What amount of wind resources did a 2010 NREL report indicate were available on US land?

   a. Enough to produce more than half of needed energy
   b. Enough to produce twice the needed energy
   c. Enough to produce 4 times the needed energy
   d. Enough to produce 9 times more than the needed energy

10. Photovoltaic systems use sunlight to _____.

    a. produce heat
    b. produce steam
    c. generate electrical power
    d. initiate fuel cell reactions

Trade Terms Crossword

74101-11 Introduction To Alternative Energy

Name: _____

Date: _____

74101-11_CW01A.EPS

Across:

3. A power generating unit used to satisfy electrical demands beyond the lowest common volume

5. An area that strives to maximize energy efficiency and use locally generated renewable energy

6. Create electric power through chemical reaction

7. Energy from sources other than fossil fuels

9. Materials based on living or recently living organic matter

10. The name of the process that creates electrical power from solar energy

14. Energy from sources that are naturally replenished

18. Electrical infrastructure with two-way communication and control

19. Devices that use power even when they appear to be off

21. The power used when one ampere flows at a voltage of one volt

22. Allows the electric meter to track power both coming and going

25. The simultaneous generation of usable electric power and heat from a single source or process; also known as cogeneration

26. Elements or metallic alloys that are capable of conducting electrical power with little or no resistance to flow

27. Energy that uses the sun to produce heat

28. Power generation of a size that is usable by and significant to utility systems

29. Simultaneous generation of usable power and heat from a single source or process; also known as CHP

Down:

1. System that transmits and distributes power

2. Solar thermal systems which can produce power 24/7

4. Requirements placed on power-generating companies and utilities regarding the implementation of power generated by renewable sources

8. The total value of all goods and services produced within the borders of a country over a given period

9. A usable fuel made from biological sources

11. Power generated in large volumes at a single location

12. The primary unit used to measure and bill the consumption of electrical power

13. Trap solar heat and radiation, causing the Earth to become warmer

15. An agreement between small power producers and utility companies

16. A power generating unit committed to handle the lowest expected electrical loads, typically running around the clock

17. One thousand watts

20. A chemical or heat process that results in changing a substance into a usable gas

23. Infrastructure to provide power from distributed sources in a localized area

24. 1,000,000 watts

Kent Peterson

Energy Supply Technical Instructor/
Training Program Coordinator
Xcel Energy, Minneapolis, MN

How did you get started in the power or energy industry?
Initially I was looking for a job to earn money for college. As I continued to rise in the ranks, taking on additional job responsibility, I decided to stick with it. Eventually I earned my B.S. through work-study and night, evening, and online courses. The work experience and the degree gave me the opportunity to sign a job transfer to our corporate training division.

What inspired you to enter the industry?
I entered the utility industry as I was looking for a job to help pay for college. A summer job transitioned into a 25-plus-year career of diverse, satisfying work.

What do you enjoy most about your job?
My job allows me to use the skills I've developed, manage my own time and resources, travel to different plants and training environments, and meet some very interesting, intelligent people from throughout the industry.

Do you think training and education are important in the power and energy industry?
Yes, training and education keep employees informed and educated on doing the right thing, the first time, safely.

How important are NCCER credentials to your career?
Very important; I would like to use these credentials, experiences, and contacts in the future for continued/additional Subject Matter Expert-type work.

How has training/construction impacted your life and your career?
Without training, there is a good chance I would still be doing shift work with a very high level of job stress. Instead, I've been in the training environment for almost 20 years and it's given me a stable, challenging career, with a very good level of income.

Would you suggest power and energy as a career to others? If so, why?
Yes; however, I would suggest that they look more into the alternative, "green" side of things. I feel that this will continue to be a career path that will provide security, challenge, and a good income for years to come, as everyone needs electricity.

How do you define craftsmanship?
Having the skill and expertise to do something with a high degree of aptitude.

Trade Terms Introduced in This Module

Alternative energy: Energy that is provided by means not related to fossil fuels. Most forms of alternative energy are considered renewable, but not all.

Base load unit: An electrical power-generating unit that has the primary mission of supporting basic power needs and typically operates on a full-time basis. A base load is considered the lowest amount of electrical power that is required to satisfy consumer needs at any time.

Biofuel: A fuel that originates from an organic, renewable source, such as ethanol or methane.

Biomass: Fuel that originates from living or recently living organic matter, generally with little or no processing before use.

Centralized power generation: The generation of electric power in large volumes at a single location to serve many consumers. Most locations fitting this description are fossil fuel, nuclear, or hydroelectric facilities.

Cogeneration: The simultaneous generation of usable electric power and heat from a single source or process; also known as combined heat and power (CHP).

Combined heat and power (CHP): The simultaneous generation of usable electric power and heat from a single source or process; also known as cogeneration.

Concentrated solar thermal (CST): The process of generating energy by concentrating the sun's power to a single point, heating liquids to create steam or high-temperature liquids. The medium can be stored to allow for power generation when solar power is not available.

Electric grid: The electrical infrastructure that transmits and distributes power across the United States.

Fuel cells: Devices that generate electrical power by using the chemical energy released by a fuel and oxidant reaction.

Gasification: Any chemical or heat process that results in changing a substance into a gas. Gasification is used to transform biomass and other solids into a useful gas, commonly known as syngas.

Greenhouse gases: Gases that contribute to the Earth's warming by trapping and reflecting solar radiation and heat back towards the planet. CO_2 and ozone are the two primary greenhouse gases.

Gross Domestic Product (GDP): The final value of all goods and services produced within the borders of a country during some period. Several methods can be used to determine the total value. Government, or public spending, is often separated from all other spending.

Interconnection agreement: A contract between a utility and a power provider that outlines the specific means of connection and how both parties will be compensated.

Kilowatt (kW): One thousand watts.

Kilowatt-hour (kWh): The primary unit used to bill for the consumption of electric power. For example, one kWh is equal to the use of one kW of power for one hour, or the use of two kW of power for 30 minutes.

Megawatt (MW): One million watts.

Microgrid: A small power grid that typically includes several forms of power generation and the primary users.

Net metering: A system of monitoring power coming into the grid as well as grid power used by a consumer. In most cases, the electric meter can turn in either direction.

Peaking load unit: An electrical power-generating unit used to satisfy electrical demands that are above the volume considered the base load. Peaking load units are operated intermittently as needed.

Phantom loads: Devices that use electrical power even when they are not turned on or actively engaged in their design function.

Photovoltaic (PV): Describes the process of transforming light energy into electrical power.

Renewable energy: Energy that comes from a source that is naturally renewed or sustained. Renewable energy sources are also considered alternative energy sources.

Renewable Portfolio Standard (RPS): Also known as a Renewable Electricity Standard at the federal level. An RPS outlines the state-required production or procurement of power generated from renewable energy sources by utility companies and other power providers.

Smart grid: A digitally enhanced grid system to allow greater control of both loads and power resources.

Solar thermal energy (STE): Energy that uses solar power to generate heat.

Superconductors: Elements or metallic alloys that are capable of conducting electrical power with little or no resistance to flow, eliminating losses. Typically, these materials function this way at a temperature near absolute zero; however, new discoveries of materials that exhibit this characteristic well above absolute zero give hope for future materials that can do so at normal temperatures.

Utility-scale: Power generation on a scale that is usable and significant to utility companies. The power may be generated by the utility's own systems or by others who sell power to the utility directly. Utility-scale power generating systems are typically listed with the DOE as such a facility.

Watt: An SI unit of power measurement equal to the power produced when a current of one ampere flows at a potential electrical difference of one volt. One watt is equal to one joule per second.

Zero energy district: An area that seeks to both reduce power consumption and generate all needed power from local alternative sources.

Additional Resources

This module presents thorough resources for task training. The following resource material is suggested for further study.

Your Role in the Green Environment, NCCER Module 70101-09. Prentice-Hall Education.
America's Energy Future: Technology and Transformation. Summary Edition. National Academies Press.
Electricity From Renewable Resources: Status, Prospects, and Impediments. National Academies Press.
US Department of Energy. www.energy.gov.
US DOE, Office of Energy Efficiency and Renewable Energy. www.eere.energy.gov.
US Energy Information Administration. www.eia.doe.gov.
National Renewable Energy Laboratory. www.nrel.gov.
North American Electric Reliability Corporation. www.nerc.com.
Center For Energy Workforce Development. www.cewd.org.
Database of State Incentives for Renewables and Efficiency. www.dsireusa.org.

Figure Credits

Courtesy of DOE/NREL, Module opener, Figures 1, 6, 13, 18–20, 27, and *Project 2: Lighting Energy Audit*

US Energy Information Administration, US Department of Energy, Figures 3, 10, and 22

C. Crane Company, Inc., http://www.geobulb.com/, SA01

US Department of Energy, Office of Energy Efficiency and Renewable Energy, Figure 4

Lawrence Berkeley National Laboratory, Figure 5

Unirac, Inc., Figures 8 and 14

©2011 Photos.com, a division of Getty Images. All rights reserved., Figures 11 and 16

Photo courtesy of Bonneville Power Administration, Figure 12

©iStockphoto.com/sharifphoto, Figure 17

©Jim Tetro Photography/Courtesy of US/DOE, SA02

Photo by Randy Montoya, Sandia National Laboratories, Figure 23

NASA, SA04

Courtesy of US Department of Energy, Figure 24

Xcel Energy, SA05

Tye Eden, SA06

Picture taken by Trayvon Leslie of Georgia Power, Figure 25

US Environmental Protection Agency, Figure 26

Lawrence Livermore National Laboratory, Figure 28

Courtesy of GreenLearning Canada Foundation, www.GreenLearning.ca, *Project 3: Building a Hydroelectric Generating Unit*

NCCER CURRICULA — USER UPDATE

NCCER makes every effort to keep its textbooks up-to-date and free of technical errors. We appreciate your help in this process. If you find an error, a typographical mistake, or an inaccuracy in NCCER's curricula, please fill out this form (or a photocopy), or complete the online form at **www.nccer.org/olf**. Be sure to include the exact module ID number, page number, a detailed description, and your recommended correction. Your input will be brought to the attention of the Authoring Team. Thank you for your assistance.

Instructors – If you have an idea for improving this textbook, or have found that additional materials were necessary to teach this module effectively, please let us know so that we may present your suggestions to the Authoring Team.

NCCER Product Development and Revision
13614 Progress Blvd., Alachua, FL 32615

Email: curriculum@nccer.org
Online: www.nccer.org/olf

❏ Trainee Guide   ❏ AIG   ❏ Exam   ❏ PowerPoints   Other _____

Craft / Level: _____   Copyright Date: _____

Module ID Number / Title: _____

Section Number(s): _____

Description: _____

_____

_____

_____

_____

Recommended Correction: _____

_____

_____

_____

_____

Your Name: _____

Address: _____

_____

Email: _____   Phone: _____

74102-11

Biomass and Biofuels

Trainees with successful module completions may be eligible for credentialing through NCCER's National Registry. To learn more, go to **www.nccer.org** or contact us at **1.888.622.3720**. Our website has information on the latest product releases and training, as well as online versions of our *Cornerstone* newsletter and Pearson's product catalog.

   Your feedback is welcome. You may email your comments to **curriculum@nccer.org,** send general comments and inquiries to **info@nccer.org**, or use the User Update form at the back of this module.

Copyright © 2011 by the National Center for Construction Education and Research (NCCER) and published by Pearson Education, Inc., publishing as Prentice Hall. All rights reserved. Manufactured in the United States of America. This publication is protected by Copyright, and permission should be obtained from NCCER prior to any prohibited reproduction, storage in a retrieval system, or transmission in any form or by any means, electronic, mechanical, photocopying, recording, or likewise. To obtain permission(s) to use material from this work, please submit a written request to NCCER Product Development, 13614 Progress Blvd., Alachua, FL 32615.

V.1 8/11

Objectives

When you have completed this module, you will be able to do the following:

1. Define biomass, identify potential sources, and describe how it is used to generate energy.
2. List the advantages and disadvantages of biomass use for energy production.
3. Describe the past, present, and future of biomass for energy.
4. Define and identify biofuels, their sources, and how they are used to generate energy.

Performance Tasks

This is a knowledge-based module; there are no performance tasks.

Trade Terms

Anaerobic digester
Biogenic
Cellulose
Convection
Cropland
Distributor plate

Metabolic
Microbes
Municipal solid waste (MSW)
Pyrolysis
Stover

Industry Recognized Credentials

If you're training through an NCCER-accredited sponsor you may be eligible for credentials from NCCER's Registry. The ID number for this module is 74102-11. Note that this module may have been used in other NCCER curricula and may apply to other level completions. Contact NCCER's Registry at 888.622.3720 or go to nccer.org for more information.

Contents

Topics to be presented in this module include:

Figures and Tables

1.0.0 INTRODUCTION

Biomass is material produced from living or recently living plants or animals and their **metabolic** byproducts. Cow manure is an example of a metabolic byproduct. Garbage is one source of biomass. However, garbage often contains non-biomass material such as plastic, glass, or metal. Wood is the primary resource for biomass energy. Crops, crop waste, and landfill gases are also sources. Metabolic processes such as photosynthesis make energy and matter available to cells.

Fossil fuel, such as coal, oil, or natural gas, is a material formed in the earth from plant or animal remains that lived in an earlier geological time. Biomass is animal or plant matter that died recently, perhaps several months ago or only yesterday.

Organic plant material contains energy absorbed from the sun. Animals absorb chemical energy from eating the plants. Energy is produced in the form of heat when biomass is burned. A fireplace is an example. Heat can be used to produce steam. Steam can be converted into mechanical energy and used to produce electricity.

Biofuels are fuels made from biomass. They can be solid, liquid, or gas. Worldwide, the primary source of fuel energy is wood. Other sources are also used, including crops, agricultural waste, and many of the components in industrial waste. Landfill gases are another source. In the United States, about 4 percent of the energy consumed comes from biomass.

Biomass is considered renewable energy because more plants can be grown, and living creatures always produce waste. The source material for biomass is constantly replenishing itself.

2.0.0 BIOMASS SOURCES AND USES

Developing countries are primary users of biomass for energy, making it the fourth largest energy source worldwide. Grassy or woody plants, dung, and garbage have been burned for heat ever since man learned to use fire. When biomass is burned, it releases about as much carbon dioxide (CO_2) as fossil fuels do, but in varying degrees. Biomass is able to somewhat balance the emissions by absorbing carbon dioxide during its growing stage.

In a process called photosynthesis, growing plants absorb sunlight and convert it into chemical energy. To convert sunlight, plants take in carbon dioxide and water. This process takes place mostly in the leaves. Plants store chemical energy as a form of sugar and use it for energy to grow

and produce seeds, fruit, and flowers. During the process, they absorb carbon dioxide and release oxygen into the atmosphere. Trees and other living plants are essential to life on Earth.

2.1.0 Biomass Sources

In today's world, with climate change and the need to reduce dependence on fossil fuels, biomass offers great potential for a clean, sustainable energy source. Biomass may significantly reduce greenhouse gas emissions. Many biomass sources are carbon neutral. This means the amount of carbon being stored by the growing plants as they take in carbon dioxide balances the carbon emissions released when the plants are used as fuel.

In the United States, many agricultural and wood-related industries use biomass for energy. For example, corn is used to make ethanol. Soybeans are a primary source of biodiesel. Using technology developed by the National Renewable Energy Laboratory (NREL), **stover** and wheat straw will soon be used (*Figure 1*).

Some biomass crops, such as fast-growing grasses and trees, are being grown on worn-out or otherwise unsuitable land that cannot be used for food crops.

74102-11_F01.EPS

Figure 1 Stover.

2.1.1 Wood

Wood is the most common source of biomass energy on earth. Much of the Third World still uses it as a primary energy source. Until the mid-1800s, wood was the main source of energy in the United States.

Because woody-stemmed shrubs and trees contain both oils and carbohydrates, they can be used to produce liquids as well as gases for energy. About 50 percent of the dry weight of wood is **cellulose**. The cellulose and starch in trees can be used for ethanol as well as for paper, cardboard, and cellophane. Burned wood can be made into charcoal, which is widely used in less technically developed countries.

Some forms of biofuel produce less harmful carbon than others, but anything that is burned emits carbon. The main forms of wood used for biomass are wood chips, sawdust, and wood pellets. Some of this wood, especially sawdust, is usually considered waste material.

Through chemical processes, wood can be converted to liquid fuel, including methanol, ethanol, and diesel.

2.1.2 Garbage

About 12 percent of the biomass energy in the United States comes from garbage. The kind of garbage found in a landfill is referred to as **municipal solid waste (MSW)**. MSW contains a large percentage of biomass in the form of food waste, garden or yard trimmings, and wood products. It also contains non-biomass materials (*Figure 2*), such as plastics, Styrofoam, synthetic materials,

TOTAL MSW GENERATION (BY MATERIAL) IN 2008

250 MILLION TONS (BEFORE RECYCLING)

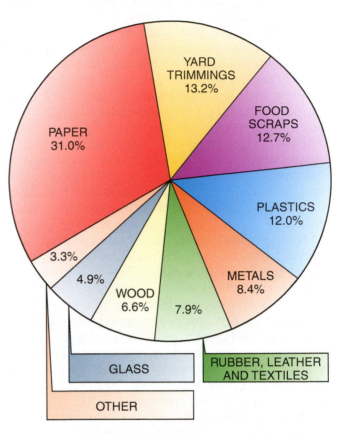

Source: U.S. Environmental Protection Agency, *Municipal Solid Waste Generation, Recycling, and Disposal in the United States: Facts and Figures for 2008* (November 2009).

74102-11_F02.EPS

Figure 2 Sources of MSW.

Are Wood Pellets the Answer?

GOING GREEN

Wood pellets burn cleaner than other wood products and have a number of advantages. As with all biomass, wood pellets are renewable. They burn the cleanest of any solid fuel. Because they produce a very small amount of creosote, pellet-burning furnaces require only exhaust vents instead of more expensive filtering and cleaning systems. Wood pellets can be mixed with coal in existing plants; adding 10 percent wood pellets to coal reduces nitrous oxide emissions by approximately 9 percent. Wood pellets are made from material normally considered waste. Untreated wood remains the natural color of wood, but torrefied wood is black.

74102-11_SA01.EPS

and metal components. Printing inks, ceramic items, and fabric dyes all contain the toxic metals lead and cadmium. Batteries are the largest source of these metals. Fluorescent lights contain mercury, another toxic metal.

Separating trash before pickup or sending it to the landfill can help with this problem. Batteries and fluorescent lights should be recycled whenever possible.

The EPA is working with manufacturers and major US retailers to promote more recycling options. Local municipal solid waste agencies have more information for their areas. Several websites offer more information.

With recycling becoming more popular, the amount of **biogenic** material in MSW has gone down. The amount of non-biogenic material in MSW has steadily increased during this time.

Every year Americans generate more waste. Today Americans produce about 4.5 pounds of garbage a day. In 1960, the amount was 2.7 pounds daily.

2.1.3 Landfill Gases

Landfills are primary sources of gases. Landfills are dumping grounds for all kinds of waste material. Much of it is organic, and decomposing organic matter creates gas. Most of this organic matter is cellulose. **Microbes** in the waste material break down the cellulose into methane. The methane slowly rises through the MSW to the surface.

Think About It

The Waste We Produce

Americans generate more than 1,600 pounds of waste a year for each person. In addition, today's lifestyle uses huge amounts of disposable material in the form of plastic, Styrofoam, and paper. Each person's garbage could take up two cubic yards of landfill a year. That is about the size of a large refrigerator box. The population of the United States is well over 300 million. That much garbage takes a great deal of space.

At the surface it enters the atmosphere as a strong greenhouse gas.

The US Department of Energy is working to reduce greenhouse gases. To support this, it awarded four ethanol companies funding to construct a commercial scale cellulosic ethanol production facility. To be eligible for the DOE funding, the facility must be able to produce up to 70 gallons of ethanol for each ton of landfill waste. A Lancaster, California, landfill alone receives an estimated 170 tons of biowaste every day. Together, two California plants operated by BlueFire are expected to produce up to 22 million gallons of ethanol from MSW per year. In addition to producing the ethanol, harmful emissions from the landfill would be significantly reduced. *Figure 3* shows how methane is recovered in a modern landfill.

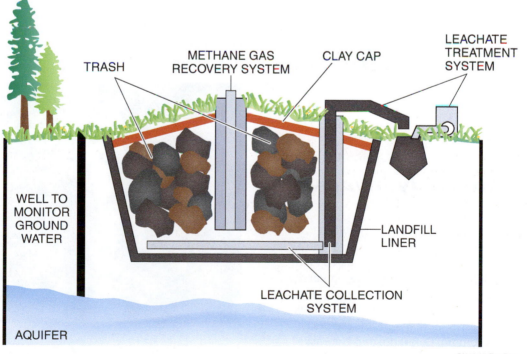

Figure 3 How methane gas is recovered from a landfill.

74102-11_F03.EPS

2.1.4 Invasive Plants

Invasive plants arrive from many places in a variety of ways. Some are ornamental plants that began as exotic garden specimens. Others were introduced to help prevent erosion or stabilize river banks. Some came accidentally in cargo and with other carriers. Some of them are damaging rivers and lakes. They can take over land areas and crowd out native species and wildlife habitats.

Recently researchers have begun to help control these invasive plants. At the same time, they hope to produce more biofuels.

Russian olive and saltcedar have infested the Northwest and pushed out cottonwoods and other native plants. They are being considered as feedstock for biofuel production.

Kudzu heads the list of invasive plants in the South. Kudzu covers an estimated seven million acres. It strangles trees and covers roads and houses or anything its path that doesn't move (*Figure 4*).

As much as 68 percent of the carbohydrate in kudzu comes from the roots. The leaves and vines hold only a small percentage. According to researchers, an acre of kudzu could yield roughly 2 to 5 tons of carbohydrate. That amount could produce about 270 gallons of ethanol. This is comparable to corn, but kudzu is extremely difficult to harvest and control. Because of those problems, it may not be worthwhile. Also, kudzu would not be a continuing biomass source; the goal is to eliminate it, not replant.

Water hyacinth is another invasive plant that holds promise for the production of bioethanol. The US Department of Agriculture (USDA) Federal Noxious Weed List includes hydrilla and water hyacinth. They are aquatic plants widely distributed in the southern United States. They are very aggressive in the Gulf Coast states. Water hyacinths (*Figure 5*) have been treated with diluted sulfuric acid and then fermented to ethanol in an anaerobic digester. The digested slurry from water hyacinths can be used as an eco-friendly fertilizer. Water hyacinths have produced about one and one-half times more gas than cow dung has. A combination of the two (water hyacinth and cow dung) is now being studied.

2.2.0 Biomass Energy Crops

Biomass energy crops are areas planted specifically for energy. Many have trees but some are grasses or woody plants with high carbohydrate content. Trees take in carbon dioxide and store it. This may make some of them carbon neutral when used for fuel. It depends on how much carbon dioxide they store, how long before they can be harvested, and the efficiency of the process for turning them into usable energy.

Energy crops are a renewable source of energy. Horticulturists and scientists are looking for high-energy, short-rotation plants. Short rotation

74102-11_F04.EPS

Figure 4 Kudzu-covered trees.

On Site

Grants to Study Invasive Plants

The Natural Resources Conservation Service, Conservation Innovation Grant program, recently awarded $1 million to the Center for Invasive Plant Management (CIPM) and Montana State University (MSU). Their project is to develop innovative ideas for managing invasive plants and work with public and private partners in Montana, North Dakota, South Dakota, Wyoming, Colorado, and Nebraska.

74102-11_F05.EPS

Figure 5 Water hyacinth.

GOING GREEN

Orange Juice Goes Green

Citrus is another promising biomass crop. Most of the peel is currently a form of waste, and Florida's fruit juice industry produces nearly five million tons of citrus waste every year. Recent technological developments have substantially reduced the cost of converting citrus peel to ethanol, making the use of this waste attractive to industry. As a side benefit, valuable peel oil is released during processing: the oil can be used as a fragrance in cleaning products, an ingredient in pelletized animal feed, and as a solvent.

74102-11_SA02.EPS

means more harvests and less time waiting for the plants to grow to harvest size.

Ash and waste products from burning biomass are generally helpful to the soil. However, the carbon emissions into the atmosphere must be controlled and reduced.

2.2.1 Productive Crops

Biomass energy crops can be a mixed blessing. Although they provide many benefits to the environment, they can also have a number of harmful effects. Land use is one factor. The need for fuel crops must be carefully balanced with the need for food crops when planning the use of land. Sometimes biomass energy crops can be established on otherwise undesirable land. Then it becomes a benefit.

Willow trees are being planted in a number of places because of their fast growth rate and high energy yield (*Figure 6*). Willows are also easily harvested with modern machinery. They grow in areas that may be unsuitable for other crops. They help to control erosion and require little maintenance. They can also be converted into a number of fuels, including wood chips, ethanol, and syngas.

Some crops are much more efficient at storing carbon dioxide than others. These crops provide a more carbon-neutral footprint. More efficient plants generally release more energy when processed. Such plants can be very valuable as energy crops.

Overall, the amount of benefit relates to how efficiently the crop can be converted to fuel. A plant that requires large amounts of resources, such as fertilizer, water, and manpower, to grow must provide a very high energy yield to be prof-

itable. Harvesting equipment and the process for conversion to usable energy are also factors. Special equipment or practices that require extra labor add to the cost.

2.2.2 Photosynthetic Efficiency

Much work has been done in recent years toward increasing the yield of major food-grain crops. The need for more productive crops has become

Figure 6 Willow crop.

74102-11_F06.EPS

the focus of research groups. Research is often done at universities and with support from the USDA and US DOE.

Photosynthesis, the process by which plants convert sunlight into energy, has only recently become popular as a research possibility. Increasing the efficiency of photosynthesis so that more energy is stored may double crop yield.

The Joint Genome Institute at the DOE reports that a fungus, *Laccaria bicolor* (*Figure 7*), has a beneficial connection with trees. It promotes faster growth. The fungus grows in North American and Eurasian forests. It has the potential to develop faster-growing trees as a biomass energy resource. It can also help the tree's ability to absorb more carbon from the air. Since carbon is stored in the starchy part of plants, this should produce higher energy yields.

Many other fungal organisms help control plant diseases by forming colonies around the roots, but they have many other promising characteristics. Many have a relationship that benefits both the plant and the fungus.

In another study, treating the plant with a particular fungus caused a large increase in root and leaf growth. The sugar content increased by 172 percent, a notable increase in photosynthetic efficiency.

2.3.0 Biomass Uses

Biomass has many uses. It can be converted into liquid fuels, called biofuels; mechanical energy for producing electricity; feedstock for biomass processing plants; and many other products. Ethanol and biodiesel are the most common biofuels produced. *Figure 8* shows an ethanol plant in West Burlington, Iowa.

Ethanol is made primarily from the starch in corn. It can be blended with gasoline, used to make alcohol, and as a type of fuel. Biodiesel is made primarily from soybeans. At present it is used to reduce harmful air emissions and increase octane. Other products are made from biomass, too.

3.0.0 Biomass Energy Production

Burning biomass directly is not the only way to release its energy. Biomass can be converted to other usable forms of energy, such as methane gas or transportation fuels. Transportation fuels include ethanol and biodiesel. The method of converting it to energy can be efficient and cause few harmful effects. It can also be as basic as burning trash in the backyard. Burning trash releases a number of greenhouse gases and does a poor job of capturing the heat energy.

Facilities that convert biomass to fuels, power, and other usable products are called biorefineries. A biorefinery is similar to a petroleum refinery in that both produce a number of fuels and products from raw material.

Biomass facilities have other impacts on the communities they serve. A 50-megawatt biomass plant burning new wood requires an estimated 70 truckloads of fuel daily.

3.1.0 Energy Flow and Trophic Levels

The troposphere is the lowest layer of the atmosphere, the one closest to the earth's surface. It is on average about 6 miles deep, depending on latitude.

74102-11_F07.EPS

Figure 7 Fungus that promotes faster growth.

74102-11_F08.EPS

Figure 8 Ethanol plant.

Greenhouse gases add to global warming by trapping some of the heat reflected by the earth in the troposphere. The troposphere also contains about 99 percent of the water vapor. Weather comes from the rising and falling temperatures and convection in the troposphere. The rising temperature from Earth's surface creates unstable air currents and contributes to more volatile weather.

The troposphere also contains the pollutants, including bad ozone. Urban smog is mainly made up of bad ozone. Ozone is a naturally occurring gas. Most of the atmosphere's ozone is in the troposphere. Ozone in the troposphere can damage the tissue of living organisms and break down some materials. These problems are caused by exposure to large concentrations of ozone. The amount of ozone fluctuates in various locations. Ozone is not dispersed evenly through the troposphere but forms large pockets in some areas.

The layer above the troposphere is called the stratosphere. The stratosphere extends from the troposphere to about 30 miles above the earth. Ozone occurs in both the troposphere and the stratosphere (*Figure 9*). The ozone layer in the stratosphere is the good ozone. It protects the earth by absorbing ultraviolet light.

In the early 1980s, technological advances allowed more sophisticated measurements of the atmosphere. Scientists became concerned that Earth's protective shield was becoming thinner and thinner over the South Pole each spring. The large thinning areas in the stratospheric ozone layer caused by chlorofluorocarbons (CFCs) are called ozone holes. These areas are not actually holes but places where the protective ozone layer is extremely thin. They are too thin to protect against penetrating ultraviolet rays. Ultraviolet light can damage the tissue of living organisms and break down some materials.

The National Aeronautics and Space Administration (NASA) captured the 1979 image shown in *Figure 10*. In 2008, the Royal Netherlands Meteorological Institute Ozone Monitoring Instrument (OMI) that flies on NASA's Aura satellite captured the same image. The images show a startling difference in the ozone layer. The ozone holes are shown in purple and dark blue.

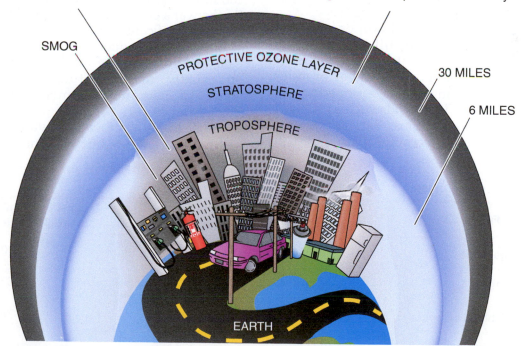

TOO MUCH HERE...

Cars, trucks, power plants, and factories all emit air pollution that forms ground-level ozone, a primary component of smog.

TOO LITTLE THERE...

Many popular consumer products like air conditioners and refrigerators involve CFCs or halons during either manufacture or use. Over time, these chemicals damage the earth's protective ozone layer.

SMOG

PROTECTIVE OZONE LAYER

STRATOSPHERE

TROPOSPHERE

30 MILES

6 MILES

EARTH

Based on *Ozone – Good Up High Bad Nearby* US Environmental Protection Agency.
www.epa.gov/air/oaqps/gooduphigh/

74102-11_F09.EPS

Figure 9 Troposphere and stratosphere.

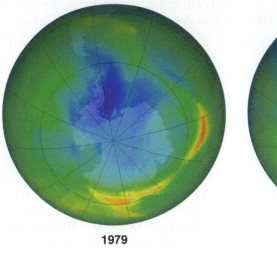

1979   2008

74102-11_F10.EPS

Figure 10 Antarctic ozone hole in troposphere: 1979 and 2008.

3.2.0 Biofuels

Any fuel made from biomass is biofuel. The name *biofuel* indicates that the fuel contains more than 80 percent renewable materials.

Some biofuels, depending on the type of plant and the process used, can produce more undesirable emissions than fossil fuels do. The production method is an important factor in reducing greenhouse gases. But even when the gases are not significantly reduced, there may still be a reason for using a particular biofuel. In addition to obtaining energy, using biofuels may help eliminate or reduce another problem, such as the enormous amount of garbage people produce.

Ethanol and biodiesel have lower energy contents than do gasoline and distillate fuel oil. Because there is less energy in ethanol and biodiesel than in gasoline or diesel, the number of miles per gallon is reduced in engines designed to run on gasoline or diesel. The blends such as E10, B2, and B5 do not have as much loss. *Table 1* compares the energy content of different fuels, using British thermal units (Btus) per gallon. The table shows both the low and high values.

3.2.1 Alcohol Fuels

Alcohol fuels are liquid chemicals that contain hydrogen, carbon, and oxygen, a combination that can be used as fuel. Man has been using alcohol as fuel for centuries.

In 1925, Henry Ford believed ethyl alcohol was the fuel of the future. Most of the automotive industry shared his opinion. By 1942, during World War II, alcohol fuel production had jumped from 100 million gallons a year to 600 million gallons. By the war's end, alcohol fuel production had become a significant factor in the war effort.

Both waste products and crops, such as sugarcane and corn, can be converted into alcohol fuels, usually ethanol. Natural gas is the primary source of methanol, but methanol can also be made from crops.

There are a number of advantages in using alcohol fuels. They burn cleaner and generate

Table 1 Biofuel Energy Content

Fuel	Btu per Gallon (Low Heating Value)	Btu per Gallon (High Heating Value)	Gallons of Gasoline Equivalent (High Heating Value)
Conventional gasoline	115,500	125,071	1.00
Fuel ethanol (E100)	76,000	84,262	0.67
E85 (74 percent blend on average)	—	94,872	0.76
Distillate fuel oil (diesel)	128,500	138,690	1.11
Biodiesel (B100)	118,296	128,520	1.03

Sugarcane is a Cool Crop

Brazil's 30-year-old sugarcane-to-ethanol project has been a success. Sugarcane fuels almost 25 percent of the country's automobiles. Research now shows another benefit—sugarcane cools the air where it grows. Its ability to reflect sunlight, draw cooling moisture from the soil, and release it into the air helps to lower the air temperature surrounding it. Carnegie Institution's Department of Global Ecology researchers say that when a sugarcane crop replaces other crops, the local climate cools.

74102-11_SA03.EPS

less carbon monoxide than gasoline does. They can be made from local crops, eliminating the need for costly transport. Alcohol fuels also have high octane. The Indianapolis 500 racetrack has been using methanol for more than twenty years. Automobiles can run on methanol and ethanol.

3.2.2 Ethanol

Ethanol, or ethyl alcohol, is a simple form of alcohol. It can be produced from any feedstock containing large quantities of sugar or easily convertible starch. These biomass feedstocks include corn, sugar cane, grains, grasses, and woods, among others. Biomass feedstocks are renewable and available, and ethanol burns much cleaner than fossil fuels.

To make ethanol, starchy corn grain is treated to extract the sugar. The sugar is fermented and then distilled. The distilled liquid, called the distillate, is then purified to produce ethanol, a pure form of alcohol. This is the same type of ethanol used in alcoholic beverages. Ethanol also yields some saleable byproducts such as corn oil and sweeteners. *Figure 11* shows the breakdown of ethanol production by use in 2010.

Engines that run on ethanol do not have the buildup of hydrocarbons, and they produce less carbon dioxide. However, ethanol contains roughly 33 percent less energy than the same amount of gasoline. The petroleum-ethanol mix E85, which is 85 percent ethanol to 15 percent gasoline, reduces vehicle mileage range by about 28 percent. The mixture most often seen at gas stations contains up to 10 percent ethanol. Although E10 reduces carbon emissions like E85, it does so

to a lesser degree; however, it has less effect on mileage than E85—a compromise.

There are also some disadvantages to producing ethanol from corn. Corn is a food supply for animals and humans, and growing it for fuel reduces the amount available for food. Government subsidies for ethanol production also raise the price of corn, making it less profitable to grow as a food crop. Another important disadvantage is that the energy required to produce ethanol from corn is almost as much as the output.

Researchers are looking for more efficient means of producing ethanol. Among their findings is switchgrass (*Figure 12*), which appears to offer a number of advantages over corn. It yields about five times more energy than goes into it. It also takes more carbon dioxide from the air than corn does.

3.2.3 Methanol

Methanol is a colorless, tasteless liquid. It is the simplest form of alcohol. It has a very faint odor and is sometimes referred to as wood alcohol. There is much interest in methanol because it can be used as transportation fuel. It has several attractive characteristics: it produces far fewer harmful emissions, such as hydrocarbons, and is much less flammable than gasoline. It produces less nitrogen oxide, another pollutant, than diesel fuels do. The cost of producing methanol is comparable to that of gasoline.

Race cars use methanol because it is a high-octane fuel with good power and acceleration characteristics.

On the downside, methanol is poisonous; for example, a few teaspoons of methanol can cause

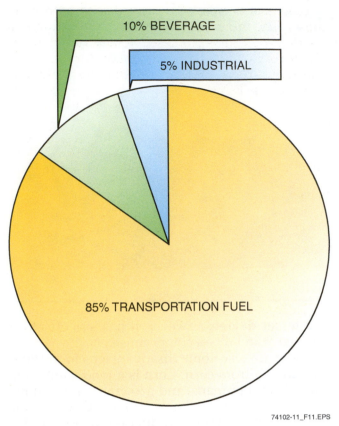

10% BEVERAGE

5% INDUSTRIAL

85% TRANSPORTATION FUEL

74102-11_F11.EPS

Figure 11 Ethanol production by use, 2010.

74102-11_F12.EPS

Figure 12 Switchgrass.

blindness and may even be fatal. Many fuels are highly toxic. Toxicity is one of the hazards associated with moonshine. However, humans can handle low concentrations, such as the amounts in cooked vegetables and artificial sweeteners. Only when the amount ingested becomes more than the body can process does it become poisonous, though the effects do not become apparent for several hours. There are antidotes if the person is treated quickly.

In the United States, methanol is mostly produced as an industrial solvent, an additive to ethanol for fuel, and for antifreeze. When it is added to ethanol, it makes it undrinkable, so producers add it to avoid paying a beverage tax on the ethanol.

Methanol can be produced from a number of sources, such as biomass, coal, and natural gas. In the United States, most methanol is made from natural gas. The cleanest plants are those that use coal to produce methanol. When coal and steam react in certain specific processes, syngas, another valuable biofuel, is formed.

3.2.4 Syngas

Syngas is synthetic gas. It is used as fuel and to help produce other chemicals. Syngas is a mixture made of hydrogen, carbon dioxide, and carbon monoxide and formed by a simple reaction between coal and steam. After processing the mixture through further chemical reactions, these components can be converted into methane.

Although it contains only 50 percent of the energy found in natural gas, syngas is still a good fuel source. To produce syngas fuel, either coal or MSW must undergo gasification. During gasification, the carbon combines with oxygen or water and produces carbon dioxide (*Figure 13*). Then the carbon dioxide combines once more with carbon, this time producing carbon monoxide.

Gasification is the primary method of obtaining syngas from biomass sources. This process changes waste feedstock into synthetic gas. Waste feedstock is placed in an oxygen-starved environment. Heat in the range of 650°C to 1,400°C (1,202°F to 2,552°F) is applied. In a direct heat process, oxygen is added to the reactor in small amounts. In an indirect process, the heat is transferred through some medium, such as hot sand.

Among the advantages of syngas is that it can be produced from MSW, which helps reduce the amount of landfill. Local production could generate jobs at the same time it helps reduce dependence on foreign fossil fuels. Syngas can be substituted for fuel oil, propane, and natural gas. It can also be used in the petrochemical industry to help make other products.

GOING GREEN

The Promise of Switchgrass

At Auburn University in Alabama, tests on switchgrass point to a number of advantages over corn. Auburn was able to harvest up to 15 tons per acre of dry biomass. That was enough for approximately 1,150 gallons of ethanol. Cutting and baling can be done by standard farming equipment; switchgrass helps control erosion and restore worn-out soil. Leftover portions of the crop can be used to make electricity.

3.2.5 Biodiesel

Soybeans are the primary feedstock for the production of biodiesel in the United States. In Europe, rapeseed and sunflower oil are the primary sources, and in Malaysia, palm oil is used. However, biodiesel can also be produced from other oils, including tallow, animal fat, and vegetable oils. It can even be produced from the grease and waste of restaurants. Biodiesel has different properties, depending on the type of feedstock used.

Biodiesel is not widely used, but because of its significant benefits to the quality of air, use is being encouraged. Recently a subsidy for users has begun stimulating interest.

74102-11_F13.EPS

Figure 13 First stage syngas compressor.

Biodiesel is most often mixed with petroleum diesel. The mixture is called B-20, indicating that about 20 percent is biodiesel. Vehicle fleets are the major users. The owners get credit for the use of alternative fuels. Automobiles that use diesel fuel can also use B-20.

3.3.0 Methods and Processes

Biomass is processed through thermochemical and biochemical methods. Thermochemical means the use of heat and chemicals. It includes pyrolysis, torrefaction, and gasification. Biochemical methods are those involving fermentation (*Figure 14*).

3.3.1 Pyrolysis

Pyrolysis is the process of decomposing biomass using heat but no oxygen. Heat at temperatures from about 400°C to 800°C (752°F to 1,472°F) is applied to biomass in a closed container called a reactor (*Figure 15*). No air or oxygen is fed to the container. The lack of oxygen prevents the biomass from combusting, or bursting into flames. Instead the product decomposes into three basic products: a liquid, a solid, and a gas. Most of the gas can be condensed into a liquid, but the pyrolysis process also produces some smaller amounts of permanent gases. Those include carbon dioxide, carbon monoxide, hydrogen, and light hydrocarbons.

Fast pyrolysis takes place in the middle temperature range over a limited amount of time, producing mostly liquid, called pyrolysis oil. Pyrolysis oil is the main product when the temperature is about 500°C (932°F). Uses of pyrolysis oil are similar to those of crude oil.

At lower temperatures and longer times, biochar is the main product. Biochar is charcoal. Only its use separates it from regular charcoal. It is highly absorbent and is used as a soil additive to improve the soil's water-holding capability as well as add nutrients.

3.3.2 Torrefaction

Torrefaction is a less intense form of pyrolysis. It usually involves temperatures in the range of 200°C to 340°C (392°F to 644°F). The purpose of biomass torrefaction is to change the feedstock properties to a more combustible form that provides much better fuel quality for use in gasification.

Any water in the biomass is removed during the process. Some of the solid material decomposes and gives off volatiles. Volatiles are the

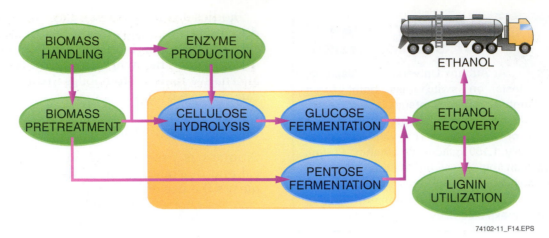

Figure 14 Bioethanol production process.

chemical elements and compounds that have low boiling points. The remaining solid material loses about 20 percent of its mass but retains most of its energy. The result is a dry, blackened material called biocoal or torrefied biomass (*Figure 16*). The torrefied biomass contains approximately 30 percent more energy than the original biomass.

After torrefaction, the biomass solids undergo densification. This is a process that compresses the material into pellets or briquettes. Densification results in a very efficient fuel product with a high energy content. Wood pellets are an example of this process.

Biomass is made of many different organic materials with many different properties. Torrefaction and densification overcome many of the differences by creating a more uniform product that has much potential as a green energy solution.

3.3.3 Gasification

The thermochemical process of gasification is a means of using heat to change fuels that contain carbon into a clean, synthetic gas called syngas. The process consists of supplying a controlled

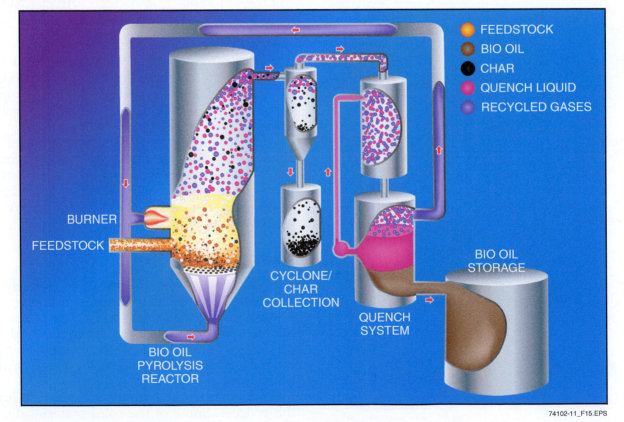

Figure 15 Pyrolysis reactor.

Think About It

Carrying Your Own Gasifier

During WWII, gasoline was needed for the war and little was available for domestic use. Nearly a million small, downdraft gasifiers were in service producing fuel, largely from wood. Some cars, trucks, and buses were equipped with individual gasifiers that used charcoal. One such vehicle was the 1941 Adler Diplomat. This process was difficult to manage and maintain and was used only as an emergency source. After the war, the idea was abandoned.

amount of air, about one-third of the oxygen needed for complete combustion. This limits the amount of burning fuel. Only a small portion is burned all the way. It provides enough heat to pyrolize, or dry, the rest of the fuel and break it down into a gas mixture. The mixture, called syngas, includes carbon monoxide, hydrogen, and methane as well as some vaporized liquids.

There are four basic types of biomass gasification methods: fixed-bed updraft, fixed-bed downdraft, bubbling fluidized-bed, and circulating fluidized-bed. The classifications are determined by the way in which the biomass in the reactor bed is supported, the flow direction of the biomass and the oxidant, and the means of supplying heat to the reactor (*Table 2*).

The oldest and simplest type of gasification reactor is the fixed-bed updraft. Biomass material is fed in at the top of the reactor. Air and oxygen feed in below the grate that supports the reacting bed. The air and oxygen rise and spread through the bed of char and biomass. Combustion is complete at the bottom of the bed. The temperature there is about 1,000°C (1,832°F). That is high enough to release the water vapor (steam) and carbon dioxide.

As the water vapor and CO_2 rise through the bed, the temperature decreases to about 750°C. The vapors change to hydrogen and carbon monoxide. The gases continue to rise and gradually dry the falling, wet biomass as it descends through increasingly higher temperatures to the bottom of the bed. There it is burned completely. Updraft gasifiers provide a simple, low-cost process, but the syngas product contains 10 to 20 percent tar. This product requires significant cleaning before it can be used in engines or turbines.

Downdraft fixed-bed gasifiers have the same operating features, but the oxidant and gas flow downward with the biomass. The air and oxygen are ignited in the top of the reactor, burning very hot. The process burns up to 99 percent of the tars, giving a much cleaner gas. Minerals, char, and ash fall to the bottom. There they are collected for disposal. The disadvantages of the downdraft process are that the biomass needs more drying before entering the reactor. Also, the syngas leaves the reactor at a very high temperature so that another heat recovery system is required.

Fluidized bed gasifiers are generally one of two types: bubbling or circulating. In a bubbling fluidized-bed, pressurized air or steam is sent through a **distributor plate** in the bottom of the reactor. The steam flows through a layer of fine particles of sand or other inert medium at high velocity. The steam causes the particles to bubble. The bubbling, or boiling, particles break up the biomass. Breaking it up causes the heat to be distributed through the feedstock.

In a circulating fluidized bed (*Figure 17*), the steam is injected at a higher velocity. This causes the feedstock particles to be trapped in the gas. The particles exit at the top of the reactor with the gas. There they are separated by a cyclone and returned to the reactor.

Table 2 Gasifier Types and Characteristics

Type	Direction of Flow		Support Type	Heat
	Oxidant	**Fuel**		
Fixed-bed updraft	Up	Down	Grate	Char combustion
Fixed-bed downdraft	Down	Down	Grate	Partial volatiles combustion
Bubbling fluidized-bed	Up	Up	None	Partial char and volatiles combustion
Circulating fluidized bed	Up	Up	None	Partial char and volatiles combustion

74102-11_F16.EPS

Figure 16 Torrefied wood and raw wood chips.

3.3.4 *Anaerobic Digestion*

Biogas is a gas mixture produced by anaerobic digestion. Biogas consists of methane and carbon dioxide. The foul-smelling gas is produced when decaying organic matter is acted on by bacteria, or microbes. This is the same process used by animal and human digestive systems.

Manure, sewage, and MSW are processed in an anaerobic digester (*Figure 18*), where microbes break down the waste material in a controlled process to produce biogas. The temperature, pH, and rate at which the material is fed into the digester are monitored and controlled to promote growth of the microbes. These microbes grow in an oxygen-free (anaerobic) environment. They eat and digest the organic matter. The result is biogas. The gas contains about 60 or 70 percent methane and 30 or 40 percent carbon dioxide. Other gases, such as hydrogen sulfide, are present in trace amounts.

Once the biogas reaches the desired level of purity, it can be used as a renewable source of natural gas. It can also be used as fuel in vehicles designed to use natural gas.

Anaerobic digestion is also the natural process that takes place in a landfill. Once a landfill contains at least one million tons of MSW and reaches 40 feet or more in depth, collecting the gas is a reasonable and beneficial process. The gas must still be treated to increase the methane and reduce the carbon dioxide and contaminants before it can be used for pipeline distribution or as transportation fuel. *Figure 19* shows a lift truck that runs on biofuel.

3.3.5 *Thermal Hydrolysis*

Hydrolysis is a process during which a molecule is separated into two parts by the addition of a

Biomass to Syngas—Solar Gasification

As part of an $18.4 million package funding twenty-one biomass research and development projects, the USDA and DOE provided a $1 million grant for a University of Colorado project to convert biomass to syngas. A team of scientists, horticulturists, and engineers is developing a rapid solar-thermal reactor for short, rapid pyrolysis, or gasification, using steam. The team found that at about 1,200°C (2,192°F), they were able to produce syngas at more than 90 percent of the original biomass. Conventional gasification results in some oxidation of the biomass, reducing yield. It also produces tars, but this rapid process prevented it.

molecule of water. Hydrolysis is used to break down a chemical compound through its reaction with water to produce other compounds. In thermal hydrolysis, the reaction is caused by heat. Because the process uses heat instead of additional chemicals, this type of hydrolysis does not produce any waste.

In pretreatment processes, thermal hydrolysis promotes more efficient conversion of biomass to bioethanol. In a research project involving discarded citrus peel, such as in a juicing plant, the result was a very high percentage of bioethanol. The citrus peel yielded ethanol, animal feed, a small fraction of essential oils, and water. The water can be used for irrigation. Thermal hydrolysis is also used with sludge, the mud-like solids produced by the treatment of sewage.

The disadvantage of thermal hydrolysis is the fouling of the heat exchangers used in the process. Any type of biomass exposed to high temperatures produces materials that gradually build in the tubes of the heat exchangers. This buildup interferes with the transfer of heat and results in more downtime and higher maintenance costs.

3.4.0 Power Generation

There are approximately 80 biomass power plants in the United States. These plants generate electricity from natural organic waste, including sugarcane, switchgrass, MSW, and wood in various combinations. A number of them operate exclusively on wood. Most of the feedstock comes from within 100 miles or less, reducing transportation costs. Some are located near pulp and paper mills and use the waste material from those facilities. Another source is logging waste, such as

GOING GREEN

Cow Power

A Vermont program called Cow Power allows customers to pay an electric premium that goes directly to the member farms that provide renewable energy. The owners of Green Mountain Dairy Farms were looking for a way to manage manure and other waste from nearly a thousand cows. They installed an anaerobic digester. They are able to sell enough electricity to power more than 300 homes with their biogas system.

An ethanol power plant in Texas uses cow manure to generate methane biogas. It saves an estimated 1,000 barrels of oil a day.

leaves and branches that might otherwise be subject to open-air burning as a means of disposal.

Wood chips are the most common feedstock, but sawdust, a waste product, is also used frequently. Sawdust, because it is light and fluffy, can create handling problems, so it is usually mixed with wood chips. Wood pellets are much more efficient and burn cleanly, but the wood must first go through another process to turn it into pellets.

3.4.1 Transport

Once the feedstock is trucked into the power plant, it is stored on site in large piles. To minimize the chances of fire as the wood ages and dries, the wood is used on a first-in, first-out basis. Bulldozers help distribute the wood in the yard.

The wood goes through a screening process to eliminate oversized or frozen chunks of wood. Magnets and metal detectors remove tramp iron, the stray metallic particles such as nails, staples, bailing wire, and other bits of metal collected with the wood.

Conveyors transport the wood to a second screening system. Wood rejects are carried to a hammerhog, a type of pulverizer with high-speed blades that grind the wood into tiny chips so they can re-enter the feedstock system.

3.4.2 Distribution

The feedstock passes another metal detector before entering the powerhouse. In the powerhouse, the wood feedstock is distributed to feeder bins. From the feeder bins, the wood goes to pneumatic distributors that convey it to the boilers. In the pneumatic distributors, high-pressure air blows the finely ground wood through large-diameter ducts to convey it from one location to another.

3.4.3 Boiler

The feedstock is deposited at a traveling gate at the bottom of a boiler, where it is dried in the combustion section (*Figure 20*). In the boiler, the wood is ignited by natural gas and burned. Natural gas is also the backup fuel. Although wood produces less slag and ash than coal, the finer wood ash can be more difficult to collect and remove. Slag is the waste material that clings to furnace walls and clogs ducts. It also collects at the bottom of the

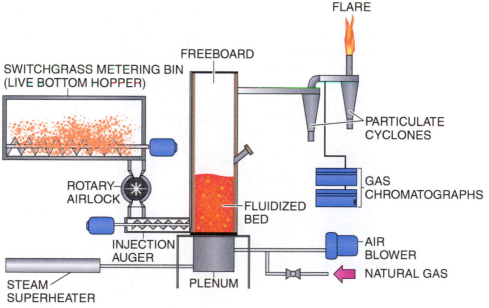

74102-11_F17.EPS

Figure 17 Circulating fluidized bed gasifier using switchgrass.

furnace and becomes black gritty material that must be periodically removed.

3.4.4 Electricity

Heat from the boiler is used to make steam. From there the power generation process is much like any other power-generating system. Stated very simply, the steam feeds a turbine that turns a generator, producing electricity. The electricity is distributed through the electrical grid for use by industry and residences.

4.0.0 ADVANTAGES AND DISADVANTAGES

Producing fuel from biomass feedstocks has many advantages. It is a renewable source; it can be produced locally, which reduces dependence on foreign sources and may create jobs. It can also reduce the enormous amount of waste products dumped in landfills. Growing plants take carbon dioxide from the air and store it.

One disadvantage is that although the source is renewable, it is not emission free. This is espe-

cially true in older processes. The emission level also varies by the type of feedstock. Some forms of methane, such as sewage methane, are highly corrosive and require more expensive equipment when used to generate electricity. In these cases, the cost of producing the fuel is high enough that it costs more than the fuel is worth.

Biomass is held back by lack of technology. Funding for research and development has been limited by the economic situation, but the need for jobs and green energy has restored interest. New projects and plants are being planned, and in many instances, construction is already under way.

4.1.0 Renewable Material

Fossil fuels are a fixed quantity and will eventually be used up. Biomass is renewable energy. More plants can be grown and more waste is being produced. Governments of many countries as well as private companies are searching for the most suitable biomass sources. They are looking for plants that grow quickly and offer a year-round harvest, as well as plants that store more

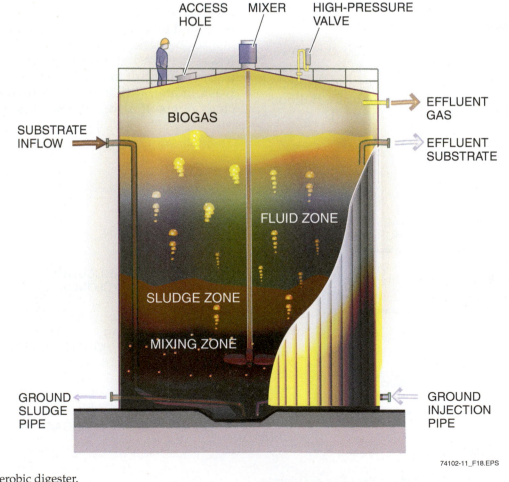

ACCESS HOLE   MIXER   HIGH-PRESSURE VALVE

BIOGAS

EFFLUENT GAS

SUBSTRATE INFLOW

EFFLUENT SUBSTRATE

FLUID ZONE

SLUDGE ZONE

MIXING ZONE

GROUND SLUDGE PIPE

GROUND INJECTION PIPE

74102-11_F18.EPS

Figure 18 Anaerobic digester.

Figure 19 Biogas truck.

74102-11_F19.EPS

carbon dioxide and more energy and are less costly to produce.

4.2.0 MSW

The ever-increasing rate at which human beings produce garbage makes MSW seem highly renewable.

Organic waste includes many things that are available in abundance: yard debris such as grass clippings, leaves, and brush; food waste and byproducts; and waste byproducts such as animal bedding. All these are excellent sources of biomass energy. Much of this is normally carted off to landfills and left to rot. There it gives off large amounts of greenhouse gases (methane). Finding a productive use for this type of waste is very important.

At this stage of technology, processing MSW is not as efficient or as carbon neutral as desired. But it does help reduce the mountains of landfill (*Figure 21*) and dispose of waste in a beneficial manner.

4.3.0 Environmental Considerations

Among the disadvantages of biomass energy is that although the source is renewable, many of the processes generate large amounts of harmful emissions. Some produce more pollution than the fossil fuels they are intended to replace. Newer technologies are making advances and finding ways to reduce this.

The need for faster growing, high-carbohydrate trees and controlled harvesting is the subject of many research projects. Willow and poplar are among the trees being used for biomass energy crops.

On the plus side, some of the crops now under consideration can help add nutrients back into worn-out soil. Others grow in swampy or arid places where ordinary plants do not survive. Plants absorb and store carbon dioxide, making the air cleaner. They provide habitat for wildlife. Trees provide shade, protection from wind, and help prevent soil erosion.

4.4.0 Costs

In some cases, the energy obtained from biomass is not enough to cover the cost of producing it. Some plants require large amounts of water and fertilizer to grow or are labor-intensive, so the profit margin is slim to nonexistent. Other biomass facilities require special equipment and frequent maintenance. The cost of purifying the product enough for use can be more than the product is worth.

However, there are many advantages. The air will be healthier as new technologies find ways to reduce greenhouse gases. Biomass energy can relieve some of the need for foreign oil. As advances are made in processes and efficiency, biomass and other alternative fuels can meet more and more energy needs.

4.5.0 Land and Resources

At present, nearly all corn and soybeans produced for biofuel feedstock are grown on prime

74102-11_F20.EPS

Figure 20 Combustion cyclone, McNeil Generating Station.

cropland in the Midwest. To continue meeting demand and achieve the goal of having biofuel make up at least 10 percent of the total gasoline consumption, the supply of feedstock must be increased. There are three possibilities. One is to increase the amount of land used, which could take more from food production. The second is to increase the energy yield of crops. Third is to replace or supplement the resource-hungry corn with other crops more appropriate for fuel production; a more appropriate crop would provide higher yield per acre at less cost. All three possibilities may be necessary.

Over the last fifty years cropland in the United States has been steadily decreasing, and now makes up about 20 percent of the total land area. As the population increases, croplands continue to give way to urban sprawl, and competition between energy and food crops is increasing. There is little chance of more cropland being added. Biofuels compete with pasture, food crops, and idle land for open space. Finding more suitable crops and increasing the energy yield is the most desirable solution.

5.0.0 PAST, PRESENT, AND FUTURE OF BIOMASS ENERGY

Biomass has been a source of energy since man learned to control and use fire. It was only in the last 150 years or so that it began to be replaced by fossil fuels. Because of climate changes and the emphasis on renewable energy, it is now the focus of a great deal of research and development.

5.1.0 Past

Biomass provided heat for warmth, cooking, mechanical energy, and most of man's needs from prehistoric times until the 1860s and the Industrial Revolution. It has been burned as a means of cooking, heating, hardening clay, and forging metal for many thousands of years.

Biomass byproducts had value from the beginning. As soon as there was fire, there was charcoal, and man no doubt discovered that it could be used to make marks. It was used in the earliest known art. The Venus of Dolní Vestonice (*Figure 22*) was found at a Paleolithic site near Brno in the Czech Republic. The ceramic statuette has been dated at 29,000 to 25,000 years BC. Many of the cave drawings, such as those found in the Chauvet caves in France were drawn with charcoal and common minerals (*Figure 23*). Tests show the Chauvet cave paintings were done in the last Ice Age, about 33,000 years ago. These are the

74102-11_F21.EPS

Figure 21 Typical landfill.

74102-11_F22.EPS

Figure 22 The Venus of Dolní Vestonice.

earliest cave drawings known. The discovery of ancient ceramics, pottery, and paintings provides ample evidence of man's early use of biomass for many things.

Coal was first used as fuel in China over 3,000 years ago, but biomass, primarily wood, continued to be much more readily available for the masses in China and most of the world. Petroleum oil in the form of seepage was called rock oil. Rock oil was also used in China thousands of

Figure 23 Lions, Chauvet cave paintings.

74102-11_F23.EPS

years ago, but it was difficult to obtain in quantity and so had limited use.

Then in 1859, Edwin Drake drilled the first successful oil well in Titusville, Pennsylvania, and fossil fuels began to overtake biomass as the primary source of energy.

5.2.0 Present

At present there is strong emphasis on developing greater energy independence and alternative fuels. Research on renewable energy sources is promising, and a number of new facilities are under construction or in the planning stages. Problems still exist in the areas of carbon-neutral and carbon-reducing technologies, but scientists, horticulturists, and engineers are working on finding solutions.

The federal government offers various help, such as tax credits, grants, and loan guarantees to encourage the use of renewable energy.

On Site

USDA Offers Energy Aid

In late 2010, the USDA announced that it would help farmers, ranchers, and rural business owners lower energy costs and develop additional sources of renewable energy through loans and grants. The program is expected to fund more than 500 renewable energy projects.

GOING GREEN

Coal Gasifiction, the Heart of the Next Generation of Clean Coal Plants?

The DOE's Office of Fossil Energy is working on a coal gasifier. Coal gasification is the process of producing coal gas from coal. It promises to be one of the cleanest ways to convert coal into electricity and other energy products. In this process, the coal is not burned directly. Instead it undergoes a chemical-heat process that breaks the coal into its basic components. In the gasifier, steam and measured amounts of oxygen break the coal molecules apart, generating chemical reactions to produce carbon monoxide, hydrogen, and other gases.

5.2.1 The DOE and USDA Positions

The US DOE and USDA support renewable energy through tax incentives and grants. The federal government supports biomass development through a variety of programs, including crop assistance programs, rural energy initiatives, and forestry research and development. However, the main focus of the DOE is on electric power generation and liquid biofuel production. Improved transportation fuels are also supported.

5.3.0 Future

By 2050, roughly 15 to 25 percent of the world's energy needs may be met by biomass sources. Depending on available land resources, energy crops such as biomass plantations could account for 20 percent to as much as 60 percent of the amount of biomass-derived energy. Reaching this goal at the present level of technology will strain natural ecosystems and habitat for endangered species. Large areas used to grow energy crops may have an unwanted effect on areas now home to a variety of plants and wildlife. Water resources must be considered.

Choosing the right crops and sources that have the most energy yield with the least impact on the environment is more important as world demand for energy continues to grow. Research, testing, and the development of new technologies is essential.

As a specific example, demand for ethanol is expected to increase proportionately as total gasoline consumption rises. Overall, gasoline consumption is projected to increase by 32 percent on

Current Research

At North Carolina State University, in cooperation with state and federal institutions, private companies, and other universities, research is taking place to analyze the best biomass for fuel and the environment. The goal is a crop that can be harvested all year instead of switchgrass and sorghum, which have three- to five-month harvest times.

In Florida, two little-known crops are being grown between crops or on fallow land. Camelina is grown for its oily seed. It can be used for both aviation fuel and animal food. The second crop is kenaf (shown in the figure). It is fast-growing biomass used mainly for oil and ethanol.

74102-11_SA04.EPS

an energy basis from 2007 to 2030. On a volume basis, consumption should be up by 34 percent by 2030. According to DOE figures, ethanol mixed with gasoline made up 4.3 percent of the total gasoline pool by volume in 2007 and should account for 7.5 percent in 2012. By 2030, the percentage will be 7.6 percent of the total.

The production of syngas is promising but not yet fully economical. Reaching an acceptable level of efficiency is still a goal. Once the syngas is produced, converting it to electric power is quite efficient, but pretreating the waste for conversion to syngas is expensive because of heavy power and oxygen consumption. It also requires an excessive amount of maintenance to keep the reactors clean. A number of experimental designs have been proposed toward solving or improving these problems.

5.3.1 Research and Development

In early 2011, the EPA announced that it is relaxing permitting requirements pertaining to biomass and other biogenic energy sources for three years. Projects directed at biomass development do not have to reduce heat-trapping pollution. This ruling is expected to jump-start biomass research and create new jobs in the industry. New designs and processes are making biomass and biofuels cleaner and more efficient. *Figure 24* shows an experimental gasification reactor.

The efficiency of burning wood byproducts is a controversial subject among scientists, but one of their findings showed that rotting wood and forest fires emit enough carbon to balance the emissions released by burning biomass. That led to

EPA's deferment of the biomass-burning requirements, giving the industry a boost.

The 50 most promising biomass development projects for the next few years include 37 US-based companies. Production is up for renewable drop-in fuels, such as green gasoline and renewable jet fuel. A biowaste-to-ethanol project in Florida, a 10-million-gallon renewable diesel project in Florida, and a 5-million-gallon biofuel project in California are among those scheduled for the near future.

5.3.2 Career Opportunities

Biomass conversion is a relatively new industry, and a focus on energy independence is driving grants and funding programs to bring more facilities on line. The need for all levels of skilled workers increases constantly as technological breakthroughs encourage new investment. The building of new fa-

74102-11_F24.EPS

Figure 24 Experimental gasification reactor.

cilities as well as additions and updates to existing plants are expected to create construction jobs.

Everything from waste collectors to harvesters to transportation suppliers are needed to collect the huge volume of waste material these plants require on a daily basis for operation. Truck drivers (*Figure 25*) and heavy equipment operators are needed to supply and manage feedstocks for the plants.

Machinery and process equipment, often new technology, needs instrumentation technicians, maintenance technicians, and operators to keep the facilities productive. Wastewater treatment plants require technicians and operators as well as maintenance and instrument technicians. As technology improves, more plants are being planned and built or expanded. Construction will require heavy equipment operators, masons, welders, ironworkers, riggers, electricians—most of the skilled trades, providing many career opportunities.

74102-11_F25.EPS

Figure 25 Tipper truck unloading wood chips.

On Site

Higher Wages in a Green Industry

A leading manufacturer of processed biomass fuel plans to build a wood pellet manufacturing facility in North Carolina, creating 53 jobs and investing $52 million. A $270,000 state grant made the project possible.

The average annual wage for the new jobs will be $38,736, not including benefits, although salaries will vary by job. In 2010, the average annual wage in the county was $27,456.

SUMMARY

Biomass is a renewable form of energy that can be converted to mechanical energy. The sources are different types of organic matter and waste. Biomass can replace or supplement fossil fuels in many applications. Although some of the processes are well-known and already in use, many more are still in development. It is an emerging technology, and the federal government supports much of the research.

Researchers are looking for alternative crops to those that are costly to grow and require special equipment to harvest. Research continues to develop methods of encouraging faster growth with greater energy value.

New facilities of all kinds are springing up, usually with government funding or support, and job opportunities in biomass production and processing are growing.

Review Questions

1. In the United States, about 4 percent of the energy consumed comes from _____.
   a. water
   b. diesel
   c. biomass
   d. gasoline

2. When the amount of carbon being stored by growing plants as they take in carbon dioxide balances the carbon emissions released when the plants are used as fuel, the plant is said to be carbon _____.
   a. rich
   b. light
   c. heavy
   d. neutral

3. Whenever possible, batteries and fluorescent lights should be _____.
   a. recycled
   b. burned
   c. crushed
   d. trashed

4. Overall the amount of benefit in selecting a crop for biomass energy relates to how efficiently the crop can be converted to _____.
   a. methane
   b. biodiesel
   c. fuel
   d. clean water

5. The lowest layer of the atmosphere, the one closest to the Earth's surface, is called the _____.
   a. stratosphere
   b. troposphere
   c. hemisphere
   d. biosphere

6. The term *biofuel* indicates that the percentage of renewable materials in a fuel is more than _____.
   a. 15 percent
   b. 45 percent
   c. 62 percent
   d. 80 percent

7. To make ethanol, corn is treated to extract the _____.
   a. oil
   b. sugar
   c. stover
   d. cellulose

8. The amount of energy in torrefied biomass exceeds the amount in the original biomass by approximately _____.
   a. 7 percent
   b. 13 percent
   c. 20 percent
   d. 30 percent

9. Biogas is produced when decaying organic matter is acted on by _____.
   a. water
   b. carbon
   c. oxygen
   d. bacteria

10. Some forms of methane, such as sewage methane, require more expensive equipment when used to generate electricity because they _____.
    a. are highly corrosive
    b. do not burn well
    c. are difficult to capture
    d. do not convert easily

11. Even though processing MSW helps reduce landfill areas, it is not yet _____.
    a. productive
    b. renewable
    c. labor intensive
    d. carbon neutral

12. On the plus side of biomass energy, some of the crops now under consideration can help _____.
    a. replenish the ozone
    b. add carbon to the soil
    c. replenish worn-out soil
    d. increase nitrogen in the atmosphere

13. The air will be healthier as new technologies find ways to reduce _____.

    a. climate change
    b. the ozone layer
    c. greenhouse gases
    d. the use of electricity

14. Coal was first used over 3,000 years ago in _____.

    a. China
    b. Greenland
    c. Central America
    d. the United States

15. One of the disadvantages of syngas production is that keeping the reactors clean requires an excessive amount of _____.

    a. carbon
    b. feedstock
    c. electricity
    d. maintenance

Name: _____

Date: _____

74102-11_CW01A.EPS

Across:

2. Heat transfer process through a gas or liquid by air circulation

5. The perforated supporting plate in the bottom of a gasifier that allows air penetration into the feedstock

6. The total area used for crops and pasture, including idle areas

7. An oxygen-free biochemical process using bacteria to decompose organic matter

9. Corn leftovers following harvest

10. Produced by living organisms

11. Referring to processes that make energy and matter available to cells

Down:

1. The exposure of biomass to high temperatures to remove volatile matter and change the remains to charcoal and oil

3. Tiny living organisms such as bacteria, funguses, and viruses

4. The kind of garbage found in a landfill

8. Fibrous carbohydrate in the walls of green plant cells that provides strength and rigidity

Sybelle Fitzgerald

Renewable Energy Project Manager Gulf Power,
Pensacola, FL

How did you get started in the power or energy industry?
I started at Gulf Power as a cooperative education student and after I graduated from Auburn University in electrical engineering, I was fortunate to be hired on full time with Gulf Power.

What inspired you to enter the industry?
I was inspired to enter the power industry by my interest in technology as well as math and science.

What do you enjoy most about your job?
The work environment is great. The people are great to work with as well. I enjoy the variety of work that is available and being able to learn new things.

Do you think training and education are important in the power and energy industry? If so, why?
Training and education are the key to being successful at Gulf Power and, for that matter, almost any other job. Training and education teach you how to perform your job correctly and safely.

How has training impacted your life and your career?
Training almost always provides advancement opportunities and increased wages. Without training, it is unlikely that I would have advanced to my current position.

Would you suggest power and energy as a career to others? If so, why?
Power and energy is a great career. It is very interesting and challenging work that provides a product that improves the lives of people around us.

How do you define craftsmanship?
A craftsman is anyone who has the skill to work with their hands to physically construct structures such as power lines, buildings, power plants, cars, and so on. Additionally, in the energy field, a good craftsman is often called upon to troubleshoot and repair damaged or non-operative equipment or structures.

Trade Terms Introduced in This Module

Anaerobic digester: Equipment used to recover methane and other byproducts by using bacteria to break down organic matter in an oxygen-free environment.

Biogenic: Produced by living organisms, such as methane.

Cellulose: The fibrous carbohydrate found in the walls of green plants cells; cellulose gives strength and rigidity to plants.

Convection: The transfer of heat through a gas or liquid by distribution of heat through currents of air circulating within a closed area.

Cropland: The total area of land used for crops, pasture, and idle land. Idle land is land that has been cultivated but is now unused.

Distributor plate: The supporting plate or grate in the bottom of a gasifier, perforated by nozzles, holes, or other pathways that allow air penetration into the feedstock resting on the plate.

Metabolic: Refers to processes that make energy and matter available to cells. Any reaction in a living organism that builds and breaks down organic molecules and produces or consumes energy in the process is metabolic.

Microbes: Very tiny living organisms, visible through a microscope, such as a bacteria, funguses, or viruses.

Municipal solid waste (MSW): The kind of garbage found in a landfill.

Pyrolysis: A process in which biomass is subjected to heat greater than 400°F to remove the volatile matter and change the remaining material to charcoal and oil. Volatile matter is the material given off as vapor or gas, but it does not include water vapor.

Stover: The leftover parts of corn after harvest, including stalks, leaves, and husks.

Additional Resources

This module presents thorough resources for task training. The following resource material is suggested for further study.

Environmental Health and Safety online. www.ehso.com
US Environmental Protection Agency. www.epa.gov

Figure Credits

Courtesy of David Nance, US Department of Agriculture, Agricultural Research Service, Module opener and SA04

Courtesy of DOE/NREL, Figures 1, 6, 12, 13, 16, 19–21, 24, and 25

Courtesy of CNFbiofuel, www.cnfbiofuel.com, SA01

US Energy Information Administration, US Department of Energy, Figures 2 and 3

Courtesy of Ted Center, US Department of Agriculture, Agricultural Research Service, Figure 4

Courtesy of Peggy Greb, US Department of Agriculture, Agricultural Research Service, Figure 5

Courtesy of Scott Bauer, US Department of Agriculture, Agricultural Research Service, SA03

Samuel Kenyon, SA02

D Vairelles (INRA), Figure 7

Photo by Tom Richard, Penn State University, Figure 8

Based on *Ozone – Good Up High Bad Nearby*, US Environmental Protection Agency, Office of Air and Radiation, www.epalgov/air/oaqps/gooduphigh/, Figure 9

NASA, Figure 10

US Department of Energy, Office of Energy Efficiency and Renewable Energy, Figure 14

Dynamotive Energy Systems, Figure 15

Dr. Robert Brown, Iowa State University, Figure 17

Simon Fraser Design, Figure 18

Courtesy of Pavel Švejnar, Figure 22

Courtesy of Miroslav Fišmeister, Figure 23

NCCER CURRICULA — USER UPDATE

NCCER makes every effort to keep its textbooks up-to-date and free of technical errors. We appreciate your help in this process. If you find an error, a typographical mistake, or an inaccuracy in NCCER's curricula, please fill out this form (or a photocopy), or complete the online form at **www.nccer.org/olf**. Be sure to include the exact module ID number, page number, a detailed description, and your recommended correction. Your input will be brought to the attention of the Authoring Team. Thank you for your assistance.

Instructors – If you have an idea for improving this textbook, or have found that additional materials were necessary to teach this module effectively, please let us know so that we may present your suggestions to the Authoring Team.

NCCER Product Development and Revision
13614 Progress Blvd., Alachua, FL 32615

Email: curriculum@nccer.org
Online: www.nccer.org/olf

❏ Trainee Guide   ❏ AIG   ❏ Exam   ❏ PowerPoints   Other _____

Craft / Level: _____   Copyright Date: _____

Module ID Number / Title: _____

Section Number(s): _____

Description: _____

_____

_____

_____

_____

Recommended Correction: _____

_____

_____

_____

_____

Your Name: _____

Address: _____

Email: _____   Phone: _____

74103-11

Nuclear Power

Trainees with successful module completions may be eligible for credentialing through NCCER's National Registry. To learn more, go to **www.nccer.org** or contact us at **1.888.622.3720**. Our website has information on the latest product releases and training, as well as online versions of our *Cornerstone* newsletter and Pearson's product catalog.

Your feedback is welcome. You may email your comments to **curriculum@nccer.org,** send general comments and inquiries to **info@nccer.org**, or use the User Update form at the back of this module.

Copyright © 2011 by the National Center for Construction Education and Research (NCCER) and published by Pearson Education, Inc., publishing as Prentice Hall. All rights reserved. Manufactured in the United States of America. This publication is protected by Copyright, and permission should be obtained from NCCER prior to any prohibited reproduction, storage in a retrieval system, or transmission in any form or by any means, electronic, mechanical, photocopying, recording, or likewise. To obtain permission(s) to use material from this work, please submit a written request to NCCER Product Development, 13614 Progress Blvd., Alachua, FL 32615.

V.1 8/11

Objectives

When you have completed this module, you will be able to do the following:

1. Define and describe nuclear power and its sources.
2. Describe and explain nuclear power development and generation.
3. List the advantages and disadvantages of nuclear power.
4. Describe the past, present, and future of nuclear energy.

Performance Tasks

This is a knowledge-based module; there are no performance tasks.

Trade Terms

Boiling water reactor (BWR)
Breeder reactor
Core
Fission
Fracturing
Fusion
Heavy water
Heavy water reactor

Light water reactor
Parabolic cooling tower
Photons
Pressurized water reactor (PWR)
Roentgens per man (rems)
Tertiary
Vitrification

Industry Recognized Credentials

If you're training through an NCCER-accredited sponsor you may be eligible for credentials from NCCER's Registry. The ID number for this module is 74103-11. Note that this module may have been used in other NCCER curricula and may apply to other level completions. Contact NCCER's Registry at 888.622.3720 or go to nccer.org for more information.

Contents

Topics to be presented in this module include:

Figures and Tables

Figures and Tables (continued)

1.0.0 INTRODUCTION

Nuclear energy is mainly used to produce electricity. In 2009, 13 to 15 percent of the world's electricity and 21 percent of US electricity was generated through nuclear energy. As oil prices rise and pollution becomes more of a concern, the demand for nuclear energy has grown. According to the US Energy Information Administration (EIA), in 2007 electricity generation from nuclear power was about 2.6 trillion kilowatt hours (kWh). It is expected to be approximately 3.6 trillion kilowatt hours in 2020 and, by 2035, it is projected to reach 4.5 trillion kilowatt hours.

Nuclear energy is generated through a method of creating heat by *fission*, or the splitting of atoms. When an atom is split, a tremendous amount of energy is released in the form of heat. In nuclear power plants, this heat is used to make steam. The steam turns a turbine, which turns a generator that produces electricity. The electricity is distributed to consumers, such as industries, businesses, and homes, by the electrical power grid.

Even though electricity is the final product, the source of the heat that produces the electricity varies. Most power plants use coal, oil, or gas to generate the necessary heat. More recently, wind, solar, and biomass power are being used, but the technology for many alternative energy sources is still being developed. The technology to use nuclear energy is already available, although researchers continue to look for more efficient and safer methods of using it.

Nuclear power plants are often located near large bodies of water because of the need for cooling water. The sight of steam billowing from the distinctive natural-draft cooling towers makes them easy to spot, especially in cold weather. Some nuclear power plants, such as the Oconee Nuclear Station in South Carolina (*Figure 1*), use a different type of tower.

2.0.0 NUCLEAR POWER AND ITS SOURCES

Nuclear energy is produced by fission, or the splitting of atoms. The most common fuel used to generate energy through fission is uranium. Uranium is a natural element found in various concentrations all over Earth, even in seawater. There is much more uranium in Earth's crust than there is gold or silver, about the same amount as tin. With the technology currently in use, the Nuclear Energy Agency (NEA) estimates that there is enough uranium to supply reactors for at least 200 years. As more efficient methods of using uranium are put into practice, this number may increase by as much as 50 percent.

74103-11_F01.EPS

Figure 1 Oconee Nuclear Station, three reactors.

2.1.0 Uranium

Uranium is a radioactive, silvery, heavy metal that is slightly softer than steel and almost twice as dense as lead. Uranium, with an atomic number of 92, is the heaviest naturally occurring element. It will dissolve in acid but is not affected by alkalis.

Most naturally occurring uranium is U-238. A second type of natural uranium is U-235, but it is relatively rare. This is the kind of uranium used in nuclear plants. It is the only natural isotope whose atoms are easily split, so it works well as fuel for a power plant. When used as a fuel, U-235 can overheat and melt if a major problem develops, but it cannot explode the way a bomb does.

To obtain more of the rarer U-235, U-238 must be enriched. First the U-238 is mined, or extracted, from whatever substance it is found in. Then it is milled and undergoes a process to enrich it so that it contains more U-235.

When the U-235 atom splits and the neutrons are released, they cause other U-235 neutrons to split off, releasing a tremendous amount of energy. This chain reaction is the basis of energy production used in nuclear power plants. The numbers following the chemical symbol U are explained by its atomic structure.

2.1.1 Atomic Structure

Nuclear energy is the energy found in the nucleus of an atom (*Figure 2*). Elements are identified by their atomic number, or proton number. The protons, which are positive, determine their properties. Uranium has 92 protons, making its atomic number 92. Elements are identified by the number of protons they have. Any element with 92 protons is uranium, no matter how many neutrons it has.

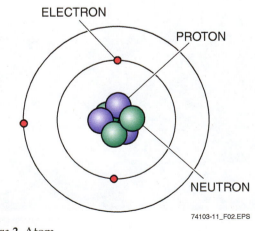

ELECTRON

PROTON

NEUTRON

74103-11_F02.EPS

Figure 2 Atom.

Most elements have the same number of neutrons as protons. The protons and neutrons in these elements balance each other, making the element neutral, but a few, such as uranium, have a different number of neutrons.

The total number of protons and neutrons in an element is its mass number, sometimes called the atomic mass number or nucleon number. The most common natural form of uranium has a mass number of 238, meaning it has 146 neutrons in addition to its 92 protons (*Figure 3*).

Some elements have variants with different numbers of neutrons. When the number of neutrons does not match the number of protons, the variations are called isotopes.

2.1.2 Radioactive Isotopes

Uranium has between 141 and 146 neutrons, creating six different isotopes. The three major isotopes, U-238, U-235, and U-234, are all radioactive and occur naturally.

Most isotopes are stable; however, radioactive isotopes, or radionuclides, have an unstable nucleus that decays radioactively. Radioactive isotopes can occur naturally or they can be induced artificially.

In radioactive decay, the radionuclide throws off particles spontaneously, that is, without colliding with another atom or particle. The rate of decay is the time it takes for half the atoms in a radioactive substance to break down, called its half-life (*Table 1*).

When radionuclides decay, they emit gamma rays or subatomic particles, sometimes both. Radioactivity occurs routinely in nature. Some materials emit radiation spontaneously. Three types of radiation are emitted: alpha particles, beta particles, and gamma rays. They are often identified in text by their Greek letters. Alpha is the Greek letter α, beta is β, and gamma is γ.

An alpha particle has the least penetrating power. It carries two units of positive electronic charge and travels only about 2 inches before it stops because of air pressure. A piece of paper or the outer layer of human skin can also stop an alpha particle.

Beta particles are a little stronger with moderate penetrating power. Each beta particle carries one electronic unit of negative charge and can be stopped by a piece of wood or a sheet of aluminum foil a few millimeters thick.

Gamma rays are high-energy photons, not particles. They have no mass and have much greater penetrating power than alpha or beta particles. It takes several centimeters of lead or about 6.6 feet of concrete to stop a gamma ray.

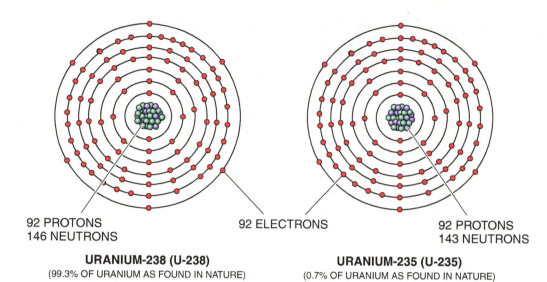

92 PROTONS
146 NEUTRONS

92 ELECTRONS

92 PROTONS
143 NEUTRONS

URANIUM-238 (U-238)
(99.3% OF URANIUM AS FOUND IN NATURE)

URANIUM-235 (U-235)
(0.7% OF URANIUM AS FOUND IN NATURE)

74103-11_F03.EPS

Figure 3 Uranium isotopes.

These radioactive emissions are the reason for the highly protective safety gear and the thick walls of the containment structure in a nuclear power plant. *Figure 4* shows the penetrating power of radioactive emissions.

2.1.3 Plutonium

Plutonium (Pu) is a solid, silvery-gray, radioactive element whose atoms can be split when bombarded with neutrons. Although it is found in only very small quantities in uranium ore, it can be produced in a nuclear reactor. Plutonium has thirteen isotopes. Of those, the most important is the isotope plutonium-239. Plutonium-239 can be used in nuclear weapons and in some nuclear reactors. Although it is almost nonexistent naturally, it can be produced in a nuclear reactor by bombarding U-238 with neutrons.

Most reactor fuel contains U-238, so during the fission process, plutonium-239 is made continuously. It is an important part of the energy cycle in a nuclear reactor. *Figure 5* shows a plutonium button in the hands of a well-protected worker. The gloves are marked with a hazardous material symbol.

2.2.0 Mining

Most uranium (U) in the United States is found in rock or gravel-like material near the earth's surface (*Figure 6*). Uranium ore is extracted through several processes.

2.2.1 Underground Mining

If the uranium is deep underground, methods similar to those used to extract other metals are used. Shafts and tunnels are dug into the earth to reach the ore. Several underground uranium mines are currently in operation, and more are planned.

The quality of the ore determines how the material is handled. If the concentration is very high, as at some Canadian mines, it must be extracted by remote equipment. In most areas of the United States, the concentrations are low, making it easier to handle as long as ventilation is good. All mines are continuously monitored for radioactivity.

Sometimes uranium is found in very low concentrations in phosphate rocks, but because the phosphate can also be mined and sold, the low-grade uranium is worth extracting.

Table 1 Uranium Radioisotopes

Radioisotope	Half-Life (Years)	Natural Uranium (Percentage)	Neutrons
U-238	4.6 billion	99.284	146
U-235	704 million	0.711	143
U-234	245,000	0.0055	142

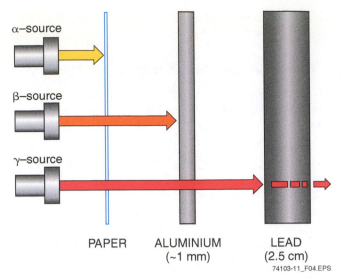

PAPER   ALUMINIUM   LEAD
(~1 mm)   (2.5 cm)

74103-11_F04.EPS

Figure 4 Penetrating power of radioactive emissions.

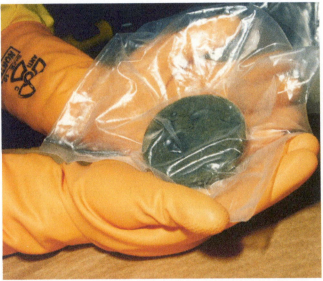

74103-11_F05.EPS

Figure 5 Plutonium button.

Uranium Ore

74103-11_F06.EPS

Figure 6 Uranium ore.

2.2.2 Open-Pit Mining

When ore is near the surface, it can be extracted by a method called open-pit mining. The area is drilled and blasted to remove large amounts of rock and expose the ore underneath. This method also produces a large volume of waste rock. The ore-bearing rock is then crushed, or milled.

Currently, uranium is not mined through open pits in the United States, but there are a number of inactive pits waiting for cleanup. The process is expensive, and some companies have gone out of business before restoring the area to the level of use available before mining. Some countries still have active open-pit mines (*Figure 7*).

2.2.3 In Situ Leach Mining

When uranium is found deep underground or in rock considered too fragile to support underground mines, a process called in situ leach mining is used. Pipes are inserted into the ore bed through holes drilled into the ground. Sometimes hydraulic or explosive **fracturing** is needed to open a path to the ore bed. A mildly caustic, or acidic, solution is then injected into the uranium bed, where it dissolves the uranium into a slurry. The slurry is pumped back to the surface operation and the uranium is extracted. The remaining water is pumped back into the ground. This method (*Figure 8*) is far safer environmentally than conventional underground mining and much less destructive than open-pit mining.

The Department of Energy's Energy Information Administration (DOE/EIA) reported that as of 2004, there were only six active uranium mines in the United States. Three of those used in situ processes. This number is expected to increase as demand generates higher prices for uranium.

In the United States, uranium ore is generally about 0.05 to 0.3 percent uranium oxide (U_3O_8). To separate the uranium oxide from the ore, the ore is milled.

GOING GREEN

Cleaning Up Open Pits

To prevent owners from walking away from open-pit mines, leaving behind wastelands and contaminated water, laws now require that companies provide cleanup money in the beginning. Mine owners must restore the land and water, including the aquifer, to the level of use before mining.

Figure 7 Open pit uranium mine, Niger.

2.2.4 Heap Leaching

In the heap leaching process, rock containing ore is placed in large piles. Chemicals are poured over the pile to dissolve the uranium, which is collected through underground drains. Currently, heap leaching is not used in the United States.

2.3.0 Milling

During the milling process, shown in *Figure 9*, the uranium is removed from the ore. The ore is fed into large crushers where it is ground into

On Site

Yellowcake Radioactivity

Neither uranium ore nor processed yellowcake emits enough radioactivity to be considered nuclear material.

uniform small particles. The waste rock, called tailings, is stored in specially designed facilities called impoundments.

After grinding, the ore is treated with acid, usually sulfuric, to dissolve the uranium. It is then purified to separate the uranium oxide, and concentrated. The product, which in early days resembled bright yellow cornmeal, is called yellowcake (*Figure 10*). Now it is more likely to be brown, but the name remains. The yellowcake is dried and stored in metal drums.

From the mill, the yellowcake is shipped to a conversion plant for further processing.

2.4.0 Converting

Uranium oxide, or yellowcake, is still not useable as fuel. The U-235 must be separated from the U-238 before it can be used in a reactor. In order to

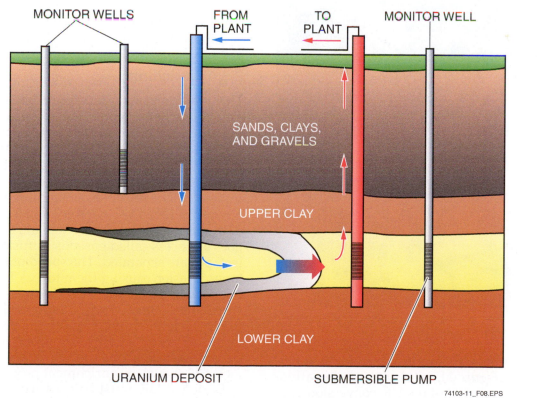

Figure 8 In situ leach mining process.

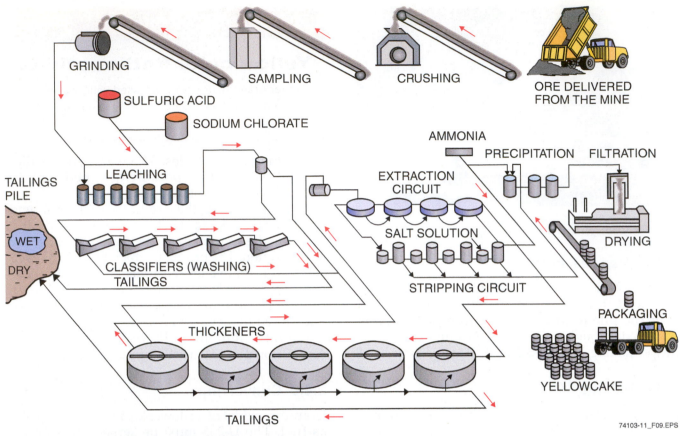

GRINDING

SAMPLING

CRUSHING

ORE DELIVERED
FROM THE MINE

SULFURIC ACID

SODIUM CHLORATE

AMMONIA

PRECIPITATION   FILTRATION

TAILINGS
PILE

LEACHING

EXTRACTION
CIRCUIT

WET

SALT SOLUTION

DRYING

DRY

CLASSIFIERS (WASHING)

TAILINGS

STRIPPING CIRCUIT

PACKAGING

THICKENERS

YELLOWCAKE

TAILINGS

74103-11_F09.EPS

Figure 9 Typical conventional uranium mill.

74103-11_F10.EPS

Figure 10 Dried yellowcake.

separate the atoms, the uranium oxide must first be converted (*Figure 11*) into uranium hexafluoride (UF_6). The product of the UF_6 conversion does not contain enriched uranium.

After separation, the gaseous UF_6 is cooled to a liquid, which drains into 14-ton cylinders. The cylinders are left to cool for approximately five days. As the UF_6 cools, it turns into a solid, which is shipped to an enrichment plant.

2.5.0 Enriching

The uranium hexafluoride produced in the conversion plant contains both U-238 and U-235. This is the type of uranium whose atoms split, or fission, easily. The amount of U-235 in uranium hexafluoride is less than 1 percent, a concentration of enriched uranium so low there is no possibility of an explosion. The purpose of enriching is to raise the amount of U-235 to between 3 and 5 percent.

In the only US enrichment plant, the UF_6 undergoes a process called gaseous diffusion, where the isotopes are separated. The speed of the U-235 isotopes is faster, causing them to leak through membrane walls before the U-238 isotopes do, allowing them to be collected while most of the U-238 isotopes are left behind. Usually the process results in a U-235 product with a concentration of 4 to 5 percent, a little above the minimum of 3 percent. *Figure 12* shows the enrichment process.

Enriched UF_6 must be converted again before it can be used as fuel. The next step is fabrication.

Figure 11 Uranium conversion plant, Blind River, Ontario.

74103-11_F11.EPS

2.6.0 Fabricating

The United States has five fuel fabrication facilities. In them, the enriched UF_6 gas is transformed into black uranium dioxide powder. The powder is compressed and shaped into small pellets. The shaped pellets are fed into a high-temperature furnace that hardens them into ceramics. After that, the pellets are polished and ground to a specified size. The pellets are about the size of a fingertip. It would take a ton of coal or about 150 gallons of oil to produce the same amount of energy as a single pellet less than 1 inch long.

Once the pellets are cooled and solidified, they can be transported to a fuel assembly plant. *Figure 13* shows a gloved hand holding an enriched uranium fuel pellet.

The ceramic fuel pellets are laid end to end in slotted forms (*Figure 14*) and fed into a fabricating machine. The fabricator inserts the polished ceramic pellets into 12-foot-long, corrosion-resistant metal alloy tubes about 1 centimeter (about 0.39 inch) in diameter. The tubes are usually made of a zirconium alloy.

Once the pellets are sealed in the rods, the rods are bundled together in a fuel assembly. The number of fuel rods in an assembly varies with the type of reactor. These rods are very heavy.

Table 2 compares a single fuel assembly for a **boiling water reactor (BWR)** to one for a **pressurized water reactor (PWR)**.

Fuel assemblies (*Figure 15*) make up the **core** of a nuclear power reactor. Most BWR reactor cores hold up to 750 fuel assemblies, while PWRs hold from 150 to 250 fuel assemblies. These numbers vary with the size of the reactor. The core is where fission takes place.

2.7.0 Fission

Nuclear fission is the splitting of the atom of a heavy element such as uranium. A slow neutron can be captured by a heavy atom, causing the atom to split, releasing more neutrons. Fast neutrons are not captured, so something called a moderator, usually water or gas, is introduced to slow the free neutrons. This increases their chances of being absorbed into the heavy atoms. Uranium-235 is the most easily fissionable material known.

To generate fission, the atoms of a radioisotope, typically U-235 in a nuclear reactor, are bombarded with free neutrons. A free neutron is

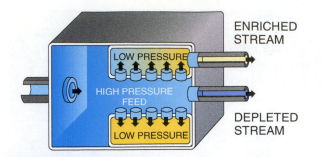

ENRICHED STREAM

LOW PRESSURE

HIGH PRESSURE FEED

DEPLETED STREAM

LOW PRESSURE

A. GAS DIFFUSION PROCESS

The gaseous diffusion process uses molecular diffusion to separate a gas from a two-gas mixture. The isotopic separation is accomplished by diffusing uranium, which has been combined with fluorine to form uranium hexafluoride (UF_6) gas, through a porous membrane (barrier) and using the different molecular velocities of the isotopes to achieve separation.

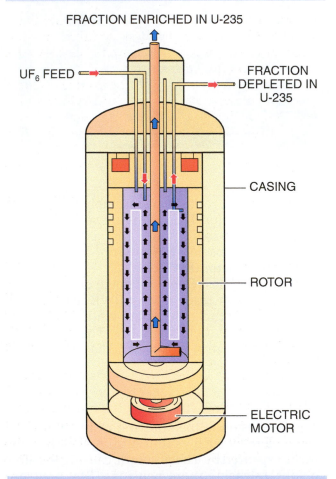

FRACTION ENRICHED IN U-235

UF_6 FEED

FRACTION DEPLETED IN U-235

CASING

ROTOR

ELECTRIC MOTOR

B. GAS CENTRIFUGE PROCESS

The gas centrifuge process uses a large number of rotating cylinders in series and parallel configurations. Gas is introduced and rotated at high speed, concentrating the component of higher molecular weight towards the outer wall of the cylinder and the lower molecular weight component toward the center. The enriched and the depleted gases are removed by scoops.

74103-11_F12.EPS

Figure 12 Enrichment process.

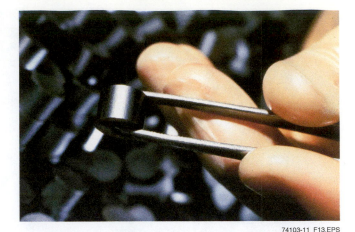

74103-11_F13.EPS

Figure 13 Uranium fuel pellet.

74103-11_F14.EPS

Figure 14 Fuel pellets aligned in trays, ready for assembly.

captured, or absorbed, by the U-235 isotope, making it unstable and causing it to split. It happens almost instantly. As soon as the free neutron is captured, the nucleus splits and throws out more new neutrons. They in turn strike other U-235 atoms, causing more atoms to split and releasing tremendous energy (*Figure 16*). This is the chain reaction that makes nuclear energy possible to use, but the reaction must be controlled. During continuous fission, massive amounts of energy are released as heat into a gas or water and used to produce steam.

Plutonium-239 is also fissionable. It does not occur naturally but is created when U-238 is bombarded with neutrons. This is the material used in nuclear weapons.

The chain reaction must be controlled to keep it from overheating. Control rods are used with the fuel assemblies to absorb free neutrons and regulate the fission process.

Table 2 Fuel Assembly Comparison

Reactor Type	Length (Feet)	Weight (Pounds)	Number of Rods
Boiling water	14.5	704	63
Pressurized water	13	1,450	264

FUEL ROD

FUEL PELLET

74103-11_F15.EPS

Figure 15 Fuel assembly with pellets, upper portion.

3.0.0 NUCLEAR POWER DEVELOPMENT AND GENERATION

Once the fuel rods are grouped into fuel assemblies, they are ready for use in a nuclear power plant to produce electricity. The nuclear process is the first part of power development, the part that makes the steam. After the steam is produced, a nuclear power plant is much like a coal- or gas-fired power plant.

The equipment for producing nuclear power is specialized with a focus on safety for both personnel and the environment. It is protected by heavy walls, and workers are required to wear personal protective equipment specially designed to protect them from radiation. Most nuclear power plants use similar equipment.

3.1.0 Major Components

The major components of a nuclear reactor are the fuel assemblies, the control rods, the coolant, and the containment vessel. Many other components are similar to those in fossil fuel-fired electrical power generation plants.

3.1.1 Control Rods

A control rod is a long, thin tube about the size of a fuel rod, but it is filled with material that absorbs free neutrons. The tubes are usually stainless steel and contain a powder or pellets. In a PWR, the rods are raised and lowered from above. *Figure 17* shows a cutaway of a Babcock and Wilcox model of a PWR at the Oconee Nuclear Station.

In a BWR, the rods are inserted from below the core. The number of rods, the degree of insertion, and the time are variable, based on the reactor design.

In some BWR designs, groups of control rods are arranged in a cruciform, or cross, shape. A cross-shaped stainless steel blade holds the control rods in position. The cruciform control rod assembly configurations are inserted into the reactor core through guide tubes that let the control rods slide in between the fuel tubes in the assemblies. *Figure 18* shows cruciform blades and *Figure 19* shows an illustration of several cruciform-shaped control rod assemblies inserted with the fuel rods.

The material in the rods is a mix of elements that absorb free neutrons. Silver, indium, cadmium, and boron are among the more common materials used, but many others including cobalt and dysprosium are also used. The materials may swell and distort the tubes over time, requiring them to be replaced. Control rod material also has a limited lifetime and eventually loses its absorption ability.

By absorbing the free neutrons, the elements in the control rods slow or even stop the chain reaction. When the control rods are removed from the assembly, fission occurs more frequently and produces more energy in the form of heat. When the reactor reaches a certain stage, the control rods

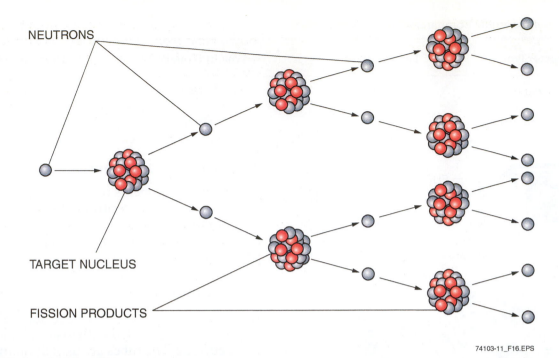

NEUTRONS

TARGET NUCLEUS

FISSION PRODUCTS

74103-11_F16.EPS

Figure 16 Nuclear chain reaction.

74103-11_F17.EPS

Figure 17 Control rods in a reactor model.

are inserted again for various reasons. Operators and monitoring equipment keep a close watch on the process so the temperature is maintained at a safe, productive level.

3.1.2 Coolant

The coolant, or moderator, is the material surrounding the fuel assemblies that transfers the heat to the generator. It also slows the free neutrons bouncing around from fission. Only slow neutrons can cause fission, so it is necessary to reduce the speed of rapidly fissioning material.

In most reactors, the moderator is plain water like that used in any plant, but in some reactors it is **heavy water**. Heavy water is water laced with a large number of molecules with deuterium atoms. Deuterium, also called heavy hydrogen, does not absorb neutrons as regular hydrogen does, so it is useful in slowing neutrons. **Heavy water reactors** are seldom used in the United States, partly because of the confusion over the radioactivity and safety of heavy water. Heavy water by itself is not radioactive. But as it absorbs the radioactive material from fission, heavy water becomes radioactive, just as plain water does.

Heat produced from fission is transferred to the coolant fluid, which can be water, gas, or liquid metal. The heat picked up by the water or liquid metal is fed to a generator to make steam, which is then transferred to a turbine. If the fluid is gas, it is fed directly into the turbine.

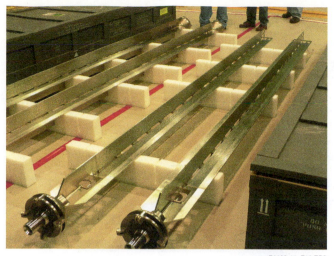

Figure 18 Cruciform blades.

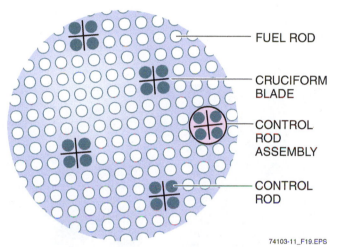

Figure 19 Cruciform control rod assemblies, viewed from the top.

Gas coolants are used occasionally. Helium is an inert gas that is sometimes used, but it cannot hold much heat and has to be circulated through the system rapidly to maintain its cooling capacity. In some cases, carbon dioxide has been used. Other coolants are being researched.

Pressurized water reactors are designed with two coolant loops because the coolant that surrounds the fuel assemblies becomes radioactive. In a PWR, the coolant remains under pressure, preventing it from expanding into steam. It flows through a steam generator in a closed loop, never coming in contact with the clean water that turns to steam. The secondary loop provides the water that turns into steam. This water is protected from the radiation so that only clean water enters the steam generator directly. By design, only excess steam produced from clean water is released to the atmosphere. Contaminated water is contained in the pressure vessel and recirculated.

Figure 20 shows the coolant flow through two loops. Heat from the primary coolant loop turns the water in the secondary loop in the steam generator into steam.

While most reactors use light water, there are a few fast reactors that do not use any type of water as a coolant.

3.1.3 Pressure Vessel

The pressure vessel is a very strong container, usually made of steel, that encloses the reactor core. The moderator, or coolant, surrounds the reactor core inside this vessel. Operating conditions inside a pressure vessel are extreme. The vessel is subjected to high temperatures and pressures as well as neutron bombardment. PWR vessels are exposed to more bombardment than BWRs, and many utilities with PWRs use special core features that lessen the number of neutrons that reach the vessel walls.

Pressure vessels are made of thick steel plates (*Figure 21*), usually a low carbon steel alloy, welded together so that walls are several inches thick, depending on the design and capacity of the reactor. One example is the San Onofre Nuclear Generating Station (SONGS) in northern San Diego County, California. Currently the plant has two operating units with pressurized water steel reactor vessels having 8-inch-thick walls. Vessels are inspected frequently to ensure their good condition.

3.1.4 Containment Vessel

A containment vessel, sometimes called secondary containment, is a concrete structure several feet thick that surrounds the pressure vessel and reactor core. The containment structures at San Onofre are 4-foot-thick reinforced concrete.

The containment vessel is designed to protect the environment and area outside the reactor from any radiation leakage in case of an accident. It will also protect the reactor against outside forces such as tornadoes or an airplane's crashing into it. *Figure 22* shows a reactor being placed in a containment vessel.

3.1.5 Steam Generator

A steam generator is a special type of heat exchanger. It is a large piece of equipment used in PWR systems. The steam generator uses the heat produced in the reactor core to change water into superheated steam.

The steam generator contains thousands (about 3,000 to 16,000) of long, thin tubes. The tubes are

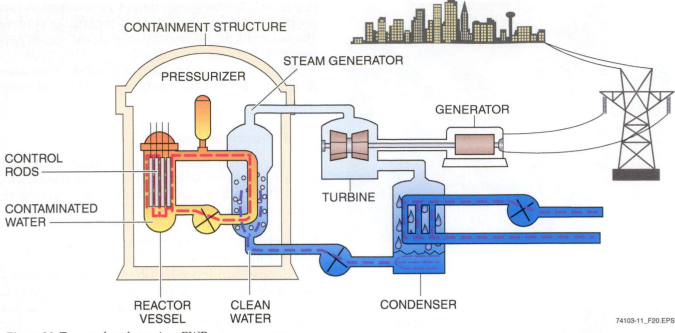

Figure 20 Two coolant loops in a PWR.

Figure 21 Nuclear reactor pressure vessel.

approximately $\frac{1}{2}$ to $\frac{3}{4}$ inch in diameter, depending on the design of the generator. Radioactive water heated by fission in the reactor is pumped through the tubes while clean water circulates around them. Water from the two systems never mixes or comes into contact.

The radioactive water in the reactor is kept under pressure to prevent it from boiling and to maintain the heat. This superheated water enters the steam generator (*Figure 23*) at the bottom and flows through the tubes. The clean water circulating outside the tubes absorbs heat through the tubes. As the temperature rises, the clean water changes to steam. The steam flows through separators to help maintain its purity. It then flows to the steam turbine, which uses it to produce mechanical energy and generate electricity.

Figure 22 Reactor placement in containment vessel.

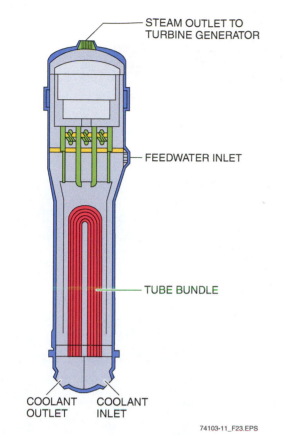

Figure 23 Heat recovery steam generator.

Much of the heat is transferred from the superheated water into the clean water, so the closed loop carries the water in the tubes back to the reactor. There, fission heats it again and the process begins once more.

The steam produced by the generator is sent to the turbine, where energy is converted to mechanical energy as it turns the turbine blades. As the energy is used, the steam cools and condenses back into water. From there it is returned to the steam generator to be converted to steam again and continue the cycle. A reactor may have up to four loops, each with its own steam generator.

3.1.6 Turbines

Steam turbines (*Figure 24*) in a nuclear power plant are very large pieces of equipment that convert the heat and pressure from steam to mechanical energy. The high-pressure steam is forced through a narrow nozzle to the turbine. It exits the nozzle at very high pressure onto the fixed blades (similar in principle to fan blades) of the turbine, causing them to turn. The spinning blades turn a shaft that goes to a generator, which produces electricity (*Figure 25*).

Turbines in power plants generally have multiple stages (*Figure 26*). The steam pressure decreases as it passes through the turbine. First is

Figure 24 Turbine in a nuclear power plant.

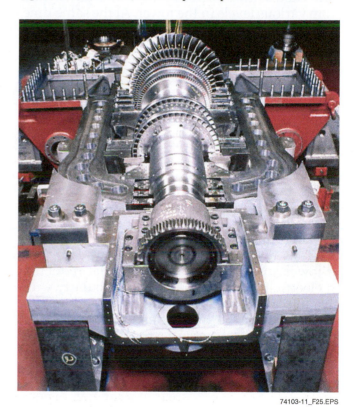

Figure 25 Turbine rotor.

the high pressure stage, then intermediate, and finally low pressure.

The spent steam leaves the turbine as a vapor. It is sent to a condenser where it is cooled to a liquid and recycled through the system.

3.1.7 Cooling Tower

In a nuclear power plant, a cooling tower is a very large structure that removes the waste heat from the cooling water in the condenser under the turbine. The cooling tower prevents hot water from being returned to the water source, such as lakes or rivers. The vapor that billows from the cooling

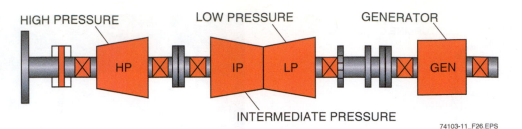

HIGH PRESSURE   LOW PRESSURE   GENERATOR

HP   IP   LP   GEN

INTERMEDIATE PRESSURE

74103-11_F26.EPS

Figure 26 Turbine stages.

tower contains only water, not any kind of radio-active material.

Two forms of cooling towers are used in nuclear power plants: mechanical draft and natural draft.

In a mechanical draft cooling tower, long pipes patterned with nozzles spray water downward from various levels in the tower. At the same time, huge fans draw the wet steam or vapor through the tower. The water droplets from the spray run over the surface of the slats, or fill, in the tower and collide with the rising vapor. Passing the falling water over the surfaces increases the contact time and helps the heat transfer. The droplets are broken into a fine spray and cooled, while the steam/vapor picks up heat.

Natural draft cooling towers operate without fans, relying on the natural upward draft of hot air through a large chimney-like structure. *Figure 27* shows an example of a **parabolic cooling tower**. Parabolic shapes are most often associated with natural draft cooling towers. Some systems use fans to supplement the air flow. This type of tower still represents a substantial reduction in electricity usage, making it particularly valuable in power plants, both coal-fired and nuclear, which need large quantities of cooling water. They have good structural strength and construction requires less material than a mechanical draft cooling tower does. Both nuclear and coal-fired power plants use natural draft cooling towers.

3.2.0 Light Water Reactors

The most common type of commercial reactor in the United States and in most of the world is the **light water reactor**, which uses ordinary water to create steam.

The heat to make the steam is produced by nuclear fission in the reactor core. There are two basic types of light water reactors: PWRs and BWRs.

3.2.1 Pressurized Water Reactors

A PWR operates on the pressure created by heating water. When water is heated, it first boils and then turns into steam. The more heat that is applied, the more the molecules spread apart and be-

74103-11_F27.EPS

Figure 27 Parabolic natural draft cooling tower.

come more active, taking up much more space. The higher the temperature, the more space they need. If the water is not allowed to expand but instead is kept under pressure, the heat is not released and the water cannot change into a vapor. As it absorbs more energy in the form of heat, it becomes superheated and the active molecules create pressure.

An example of this is boiling water in a pot with a lid. As the water heats, the molecules become more active, and the lid rattles or lifts, allowing steam to escape. If the lid is held down, the pressure will increase as long as the water temperature rises but is not given room to expand. If the pressure is relieved suddenly, the superheated liquid will burst out in a flash, turning instantly to vapor. *Figure 28* shows how the molecules behave when heated and how the water expands as it boils. This happens when the lid on a pot of boiling liquid is lifted.

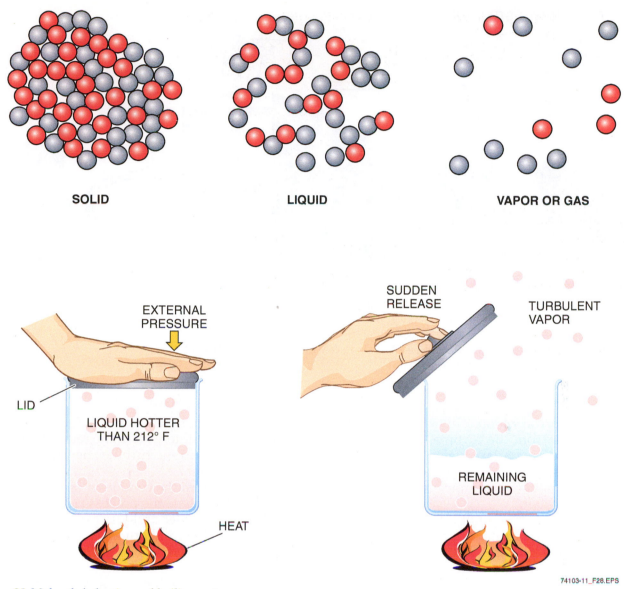

SOLID **LIQUID** **VAPOR OR GAS**

EXTERNAL
PRESSURE

LID

LIQUID HOTTER
THAN 212° F

HEAT

SUDDEN
RELEASE

TURBULENT
VAPOR

REMAINING
LIQUID

74103-11_F28.EPS

Figure 28 Molecule behavior and boiling water.

Water is pumped into the pressure vessel to the reactor through a closed loop. A closed loop means it circulates continuously through sealed pipes and vessels and cannot mix with the water from other systems.

The water (called a coolant) surrounding the fuel assemblies has a dual purpose. It moderates the action of the fissioning neutrons and prevents the reactor core from overheating. The slow neutrons are needed to strike and split the U-235 isotopes. The continuous neutron bombardment maintains the chain reaction that supports nuclear power plants. In doing this, the water picks up some radioactivity from the neutrons it absorbs, but it is never released to the atmosphere, allowed to mix with other water systems in the plant, or come in contact with any personnel. It is one of the stringent safety measures employed in nuclear power plants.

When fission occurs in the reactor, releasing tremendous amounts of energy into the water in the form of heat, the water temperature rises far above the temperature of steam. In a PWR the water is kept under pressure and not allowed to expand, so it gets hotter and hotter but stays in its liquid state, creating enormous pressure.

In a PWR, the superheated water in the primary cooling loop flows through a special heat exchanger called a steam generator that boils water from a second loop, creating the steam that feeds the turbo-generator. The primary and secondary loops of a PWR keep the radioactivity isolated so that only clean steam is circulated through the turbine. This helps to minimize maintenance costs and radiation exposure to the environment and to the plant personnel.

Pressurized water reactors have three separate water systems, shown in *Figure 29*. The third loop

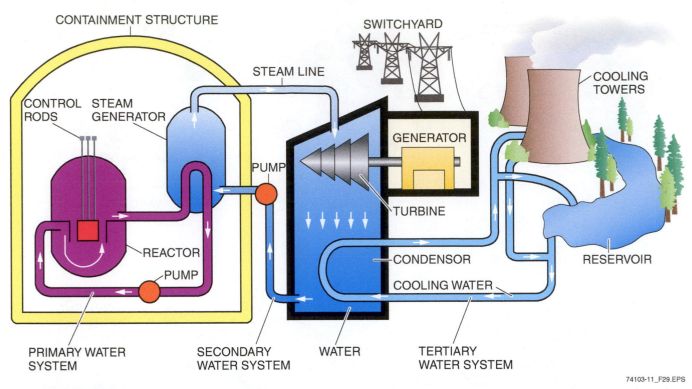

Figure 29 PWR water systems.

is called the **tertiary** loop. This is water used to cool the steam from the turbine. Tertiary water usually comes from a large lake or river, sometimes the ocean, close to the plant. The need for cooling water is why most nuclear power plants are built near water.

The reactor core must not be allowed to overheat, and there are multiple systems in place to ensure that this does not happen. Pumps powered by diesel generators are in place to supply emergency cooling water. Diesel generators operate independently and do not use power from the main electrical system.

3.2.2 Boiling Water Reactors

In a BWR, fission takes place in a reactor pressure vessel. Water is pumped into the vessel, surrounding the fuel assemblies (*Figure 30*). As fission occurs, the water boils. It slows the fast-moving, fissioning neutrons and keeps the reactor from overheating. The boiling water picks up radioactive isotopes from the fission process. It is drawn up through a separator in the top of the pressure vessel, where the water is separated from the vapor.

From the separator in the top of the reactor, the vapor, or steam, is fed directly to the turbine. The energy from the steam causes the turbine blades to rotate, turning a shaft to a generator. The generator produces electricity. The now cooling steam exits the turbine and enters a condenser where

cooling water pumped from a large body of water condenses it into water. The radioactive water from the steam in the turbines never mixes with the cooling water.

The water from the condenser is pumped back to the reactor, where the process is repeated.

BWRs are simpler and less expensive to build than PWRs. Safety in both systems is strictly regulated and enforced.

3.3.0 Heavy Water Reactors

Heavy water is water laced with deuterium, making it about 10 percent heavier than plain water. Deuterium, also called heavy hydrogen, is a stable isotope of hydrogen that occurs naturally in the oceans. With slight alterations, both PWR and BWR designs can be used with heavy water. Heavy water by itself is not radioactive; like light water, it becomes radioactive when it begins picking up radioactive neutrons in the fission process. However, it absorbs fewer neutrons than light water does.

Heavy water does not slow neutrons as much as plain water, but neither does it absorb nearly as many neutrons. The lower absorption rate allows the use of U-238, or natural uranium, but requires a slightly different configuration, such as more space between control rods, in the reactor core.

The disadvantage is the expensive vacuum distillation process required to enrich the water, or

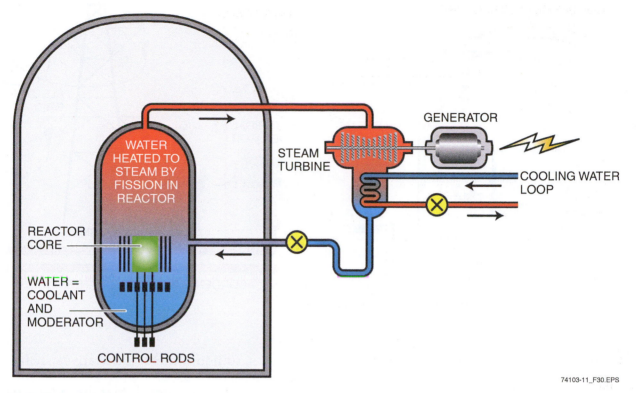

Figure 30 Boiling water reactor.

moderator, instead of the fuel. The initial cost is about 20 percent higher than that of a light water reactor. However, once the heavy water is produced, the cost of refueling the reactor with lower-grade uranium is considerably less.

The best-known heavy water reactor is Canadian (*Figure 31*). The full name is CANada Deuterium Uranium, using the trademark CANDU®. At present, all Canadian reactors are CANDU reactors. In the United States and Europe, nuclear units typically use light water reactors.

In the present state of technology, comparing the amount of uranium it takes to produce one megawatt of electricity, a CANDU reactor uses about 15 percent less than a PWR. Because uranium for a CANDU does not have to be enriched, the fuel is more widely available and enrichment facilities are not required. These heavy water reactors can even use spent fuel from other reactors, significantly reducing the need for waste disposal.

3.4.0 Breeder Reactors

A **breeder reactor** is one that creates more fissionable material than it uses. Breeder reactors use approximately 3 percent as much fuel as a traditional reactor does. The production of more fuel than is consumed is the basis of a breeder reactor. A breeder reactor needs to be operated with fast neutrons.

By encasing the reactor core in a uranium blanket, the highly fissionable plutonium-239 byproduct of fissioning is captured in the blanket instead of being absorbed into the water.

The rate at which new fissile material is produced is the breeding ratio. Thus far, the upper limit has been 1:8, or eight times more than it consumes. When uranium-238 is bombarded with neutrons one of the products is plutonium. The breeding ratio determines the amount of plutonium.

The biggest advantage of a breeder reactor is that after the initial amount of fuel is delivered, the reactor requires only occasional refueling with natural, or unenriched, uranium. There is no need for uranium enrichment plants to support a breeder reactor.

Among the disadvantages is that the high-grade plutonium produced in the reactors could make security more difficult. However, there is a simple solution: the addition of very small amounts of elements such as curium or neptunium to the plutonium during preprocessing makes the plutonium almost impossible to use in explosives.

Significant research into breeder reactors is continuing in France, India, Japan, China, and Russia.

There are several variations on breeder reactors, the best known being fast breeder reactors and thorium reactors.

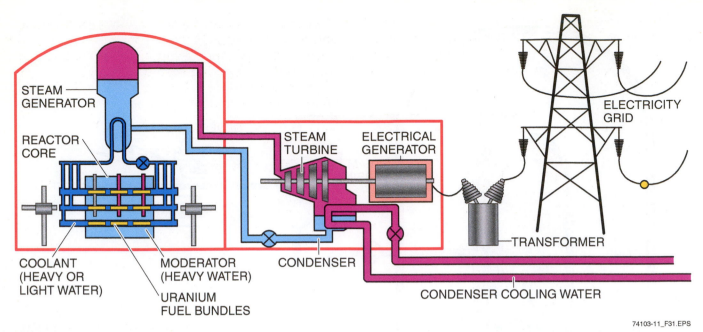

Figure 31 CANDU reactor.

3.4.1 *Fast Breeder Reactors*

In a fast breeder reactor (*Figure 32*), media other than water must be used as the moderator. Water slows the neutrons, preventing the fast fissioning process, so liquid metal alternatives such as sodium have been used.

In the liquid metal fast breeder reactor (LMFBR), a 1.4 breeding ratio is desired, but so far the process has only reached about 1.2.

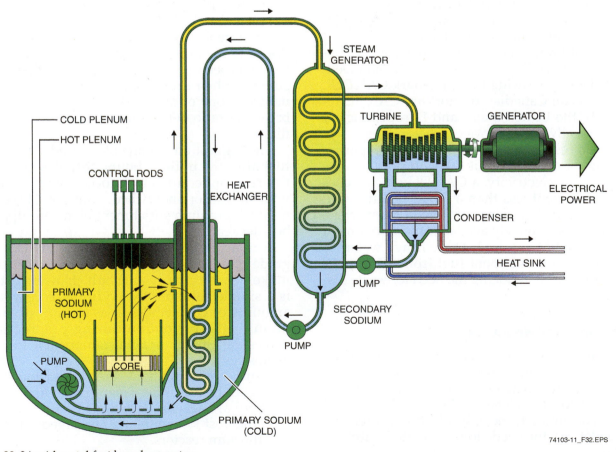

Figure 32 Liquid metal fast breeder reactor.

Three of the next generation of reactors under consideration by the US Nuclear Regulatory Commission (NRC) are fast breeder reactors. India is developing both uranium and thorium fuel for its FBR technology.

3.4.2 Thorium Reactors

Another type of breeder reactor uses thorium (*Figure 33*), a fissionable element. Thorium (Th) is a slightly radioactive silver element with the atomic number 90. On the periodic table, it falls at the bottom with the heavy elements uranium and plutonium. It is three or four times more common in Earth's crust than uranium.

Thorium is widely available throughout the world and can be extracted economically. The Thorium Energy Alliance estimates there is enough in the United States alone to meet energy needs for more than a thousand years.

India, which has large natural deposits of thorium, is actively researching thorium reactor designs. In the United States, a thorium reactor moderated by molten salt was operated successfully at Oak Ridge National Laboratory in the 1960s but was cancelled for lack of funding.

Thorium nuclear reactors cannot melt down, and they offer a relatively inexpensive way to produce electricity. Another attractive feature is that they can burn existing high-grade nuclear waste. These reactors have excellent potential for green energy and are generally supported by the environmental community.

Thorium has other advantages: its half-life is much lower than that of uranium, meaning nuclear waste from thorium decreases in radioactivity much more quickly than waste from uranium. Thorium is extremely difficult to use in bomb making.

Leading researchers believe thorium molten salt reactors are of major importance in reducing the need for fossil fuels. Although more work needs to be done, these reactors compare favorably to other power generation sources in operating costs, safety, and efficiency.

Early nuclear power plants are now referred to as Generation I. Current nuclear power plants are considered Generation II and III designs. While they serve many areas with secure, low-cost electricity, research into improved designs continues in both government and private sectors. The US Department of Energy's Office of Nuclear Energy

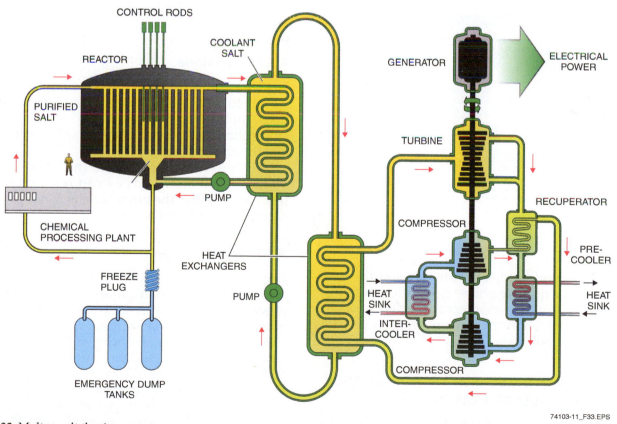

74103-11_F33.EPS

Figure 33 Molten salt thorium reactor.

GOING GREEN

Small Reactors in the Future

National laboratories at Lawrence Livermore, Argonne, and Los Alamos Argonne are working jointly on a self-contained nuclear reactor called SSTAR (small, sealed, transportable, autonomous reactor). The fast breeder reactor will produce from 10 to 100 megawatts of electricity. Shown below is a graphic representation of the SSTAR.

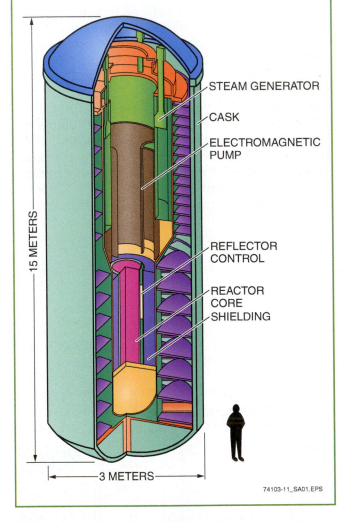

STEAM GENERATOR

CASK

ELECTROMAGNETIC PUMP

REFLECTOR CONTROL

REACTOR CORE

SHIELDING

15 METERS

3 METERS

74103-11_SA01.EPS

is part of a worldwide effort to develop the next generation of nuclear power generation systems. This research and development effort is known as Generation IV. Thorium reactors are now considered Generation IV reactors.

4.0.0 ADVANTAGES AND DISADVANTAGES OF NUCLEAR POWER

Nuclear Energy Institute (NEI) figures show that if all nuclear power plants worldwide used fos-

sil fuels, together they would add 2 billion metric tons of carbon dioxide to the atmosphere every year. But as long as they are working as designed, nuclear power plants release almost no carbon dioxide. And a single nuclear power plant releases less radioactivity than a coal-fired power plant.

On the other hand, wastes from uranium mining and spent fuel are problems. No one wants a nuclear waste repository next door. Nuclear wastes require special cleanup and handling and must be stored in well-contained, secure locations. If not properly dealt with, they are certainly hazardous.

Unfortunately, it is often difficult to separate facts from emotionally charged exaggerations on both sides of the controversy. There have been many distortions both for and against the use of nuclear energy, but the record shows far fewer deaths or illnesses from nuclear power plant operations than from fossil fuel-fired power plants.

4.1.0 Environmental Impact

A nuclear power plant itself has little environmental impact other than the normal footprint for a large industrial facility. However, some mining methods are very destructive to the land and water. The main problem is nuclear wastes, both from mining and from spent fuel. Nuclear wastes are the subject of a lot of research, and improvements continue to emerge.

4.1.1 Emissions

Nuclear power does not produce any significant wastes that are released to the environment and does not add to the carbon dioxide levels in the atmosphere. A nuclear power plant does not pollute the environment.

Uranium mine tailings are the largest source of nuclear pollutants. They can release radioactive isotopes into the environment if not properly cleaned up and contained.

4.1.2 Waste Stream

The three main kinds of radioactive wastes are spent fuel, which is high-level waste; low-level wastes such as gloves, tools, and clothing; and uranium tailings from mines.

The two methods of storing spent fuel after removal from the reactor core are spent fuel pools and dry cask storage. *Figure 34* shows a cutaway view of a dry cask, or canister.

According to the NRC, spent fuel waste management and disposal is funded from the time it is generated. Generating enough electricity for

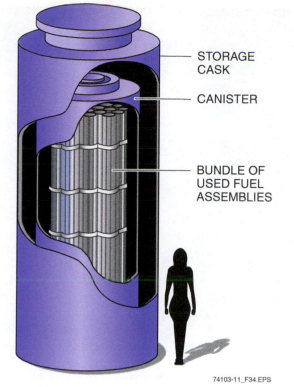

STORAGE CASK

CANISTER

BUNDLE OF USED FUEL ASSEMBLIES

74103-11_F34.EPS

Figure 34 Dry canister.

about 1,000 people (in developed countries) in a year creates about 66 pounds of spent fuel.

In 40 years, the amount of used nuclear fuel generated by nuclear power plants comes to about 62,500 metric tons. If all the spent assemblies were stacked end-to-end and side-by-side in a football field, the pile would be only about seven yards (21 feet) deep.

Spent fuel consists of the fuel rods used in the reactor core. After long use, the accessible fissionable uranium is reduced below the point of use. The rods must be reprocessed. The uranium and plutonium are chemically separated from the spent fuel and then recovered. These elements can be recycled and used in breeder reactors or for other purposes, such as instruments and gauges or medical devices. Security is emphasized and the materials are tracked and regulated.

Many nuclear weapons are being dismantled worldwide, and the uranium and plutonium are being reused as fuel in power plants. The procedures and requirements are well established and safe. The United States buys and reuses significant quantities from the countries of the former Soviet Union.

4.1.3 Half-Life and Hazards

Half-life is the time it takes for half the atoms in a radioactive element to break down. The long

half-life of U-238 (over 4 billion years) and U-235 (702 million years) is one of the problems with the waste products. Although other radioactive materials used in nuclear fuel may be much greater in quantity, their short half-life makes them easier to handle and dispose of.

If not handled properly, the radioactivity can enter streams and waterways or contaminate the land, including plants that grow and animals that graze there. Humans can suffer radiation poisoning as well. However, with proper protective equipment, waste that is contained and sealed according to regulations can be handled safely.

4.1.4 Disposal

Disposal of the spent rods from a nuclear reactor is strictly regulated. Each reactor site in the country has a spent fuel pool designed to keep radioactive waste safe for a number of years. The spent rods are carried along the bottom of water canals to ensure that neither personnel nor the environment is exposed to radiation. From the canals, the spent fuel rods are stored in pools (*Figure 35*) under a minimum of 20 feet of water. The water shields anyone in the area from the radioactive waste.

Every 12 to 18 months about 25 to 33 percent of the total fuel is removed from the reactor and replaced with fresh fuel.

The waste heat that comes from the cooling water in a PWR-type facility is the white vapor, or steam, that rises from the cooling towers. It is not radioactive.

The safety record for storage of spent fuel waste is good. The Nuclear Waste Policy Act of 1982 states that the Department of Energy is responsible for all high-level radioactive waste disposal systems. Waste must be securely isolated until any radioactivity has decayed to a level safe for humans and the environment. According to the NRC, current plans

74103-11_F35.EPS

Figure 35 Spent fuel pool.

call for the ultimate disposal of the wastes in solid form in licensed deep, stable geologic structures.

The site selected for the national nuclear waste repository is Yucca Mountain (*Figure 36*) in Nevada. Due to delays, spent fuel is currently stored at a number of sites across the United States.

Long-term storage requires that the radioactive waste be contained in a form that will neither decay nor give off radiation for very long periods of time. One way this is being done is through **vitrification**, a process in which the waste is bonded with a special form of very hard glass. Vitrified glass is highly resistant to degradation. The molten glass mix is poured into stainless steel containers, cooled, and sealed (*Figure 37*). The containers are then stored, usually in a specially designed vault deep underground.

4.1.5 Transportation

Part of the NRC's responsibilities includes overseeing the shipment of nuclear waste. Because significant amounts of radiation and heat remain in spent fuel assemblies after they have been removed from the reactor, NRC regulations require that spent fuel be shipped in specially designed casks or containers. The containers prevent radioactivity leakage and help to reduce the heat.

All shipments must be transported over NRC-approved highway routes for transport of spent nuclear fuel. Shippers must inform and coordinate with local law enforcement agencies while in transit, and armed escorts are required in heavily populated areas.

Information regarding shipment time and date is considered sensitive and not made public to protect against any threatening acts that could result in harm.

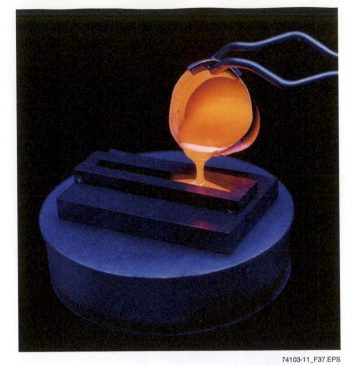

74103-11_F37.EPS

Figure 37 Glass vitrification process.

For more than 30 years, spent fuels have been shipped across the United States. None of the thousands of shipments have experienced radioactive releases. There has been no harm to the public or to the environment.

4.2.0 Cost

While total generating costs for a nuclear power plant are much lower than for coal or gas-fired power plants, the startup costs for nuclear plants are greater than the capital costs for a gas-fired plant.

Safety features required for protection of the public and the environment increase the cost of the facility and the equipment, but the actual operating cost is less. The costs of decommissioning and waste disposal are included in the capital costs.

74103-11_F36.EPS

Figure 36 Yucca Mountain.

On Site

Transporting Uranium Ore

Colorado is the site of many mining activities, including uranium mines. By Colorado state statute, uranium ore and processed yellowcake, used to make fuel rods for nuclear reactors, are considered mere hazardous materials and therefore not limited to transportation along the state's designated nuclear materials routes.

4.2.1 Life of Unit

At present the life expectancy of a nuclear power plant is up to about 60 years. The NRC grants operating licenses for 10-year periods, although plants undergo constant safety inspections. They must be maintained in good working order. Most currently operating nuclear power plants are now expected to apply for 20-year license extensions.

Over the last 25 years, the NRC has conducted many studies on extending the life of these plants and whether it would be justified economically. The answer has been overwhelmingly positive. Safety is the most important consideration. Updates and modifications may be needed or economically advantageous, but the plants are now considered to have a much longer life expectancy.

Multiplying currently available nuclear power by three could make a significant difference in reducing or slowing the effects of climate change. It is estimated by that carbon emissions could be reduced by one to two billion tons every year.

4.2.2 High Volume to Power Ratio

Even a very small amount of enriched uranium provides a tremendous amount of energy. One uranium pellet about the size of a child's marble holds the same amount of energy as 17,000 cubic feet of natural gas, 149 gallons of oil, or 1,780 pounds of coal.

The NEI published 2008 figures (*Figure 38*) for the United States regarding the cost of electricity production.

> **NOTE**
>
> This chart refers to fuel plus operation and maintenance costs only. Capital costs are not included because of differences between locations and ages of the facilities.

4.3.0 Proven Technology

Among the advantages of nuclear energy is that the technology has already been developed and is available now. In the United States, 104 nuclear power reactors are currently in operation, supporting energy needs safely, efficiently, and at low cost.

4.3.1 Sustainability or Renewability

While nuclear fuel is not renewable in the traditional sense—there is only so much uranium or thorium in Earth's crust—breeder reactors can produce an increasing volume of useable material

for a very long time. As research continues, the efficiency of breeder and fast breeder reactors is expected to improve, and the available fuel material will last for many hundreds of years.

4.3.2 Radiation Hazards and Procedures

Radiation occurs naturally everywhere on Earth through many sources. The annual percentage generated by nuclear power is very, very small, well under 1 percent of the total, normal amount a person receives. A passenger on a single cross-country airplane flight receives more. Roughly 1 percent of all cancers are caused by radiation, and the risk from nuclear technology may increase the chances by 0.002 percent or reduce life expectancy by one hour. Other forms of electricity generation reduce life expectancy by a much greater percentage.

The amount of radiation received by humans is measured in **roentgens per man (rems)**. Everyone who works with or around nuclear energy is required to wear a small device called a dosimeter (*Figure 39*), a small device that measures cumulative rem exposure (keeps track of the total). There are many different forms, but most are about the size of an ID badge and worn in a prominent place on the body, usually the chest or waist. Depending on the type of work they perform, some people may wear a ring dosimeter under gloves.

NRC statistics show the approximate amount of radiation the average person receives in a year from various sources (*Table 3*).

4.3.3 Site and Materials Security

The National Nuclear Security Administration (NNSA) is responsible for the security of nuclear sites and materials in the United States. Security programs for NNSA are developed and implemented by the Office of Defense Nuclear Security (DNS). Areas of responsibility include support security operations, technical support, and resources.

Current nuclear security regulations are designed to protect against intentional damage from several types of threats called the Design Basis Threat (DBT) for nuclear reactors. DBT is described in detail in *Title 10, Section 73.1(a)*, of the *Code of Federal Regulations [10 CFR 73.1(a)]*.

The NRC regularly reviews and revises DBTs as necessary to respond to developments in the threat environment. The DBT takes into account the 12 primary factors listed in the *Energy Policy*

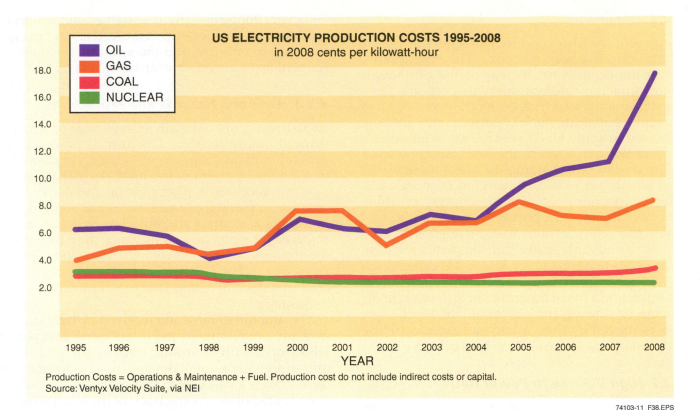

US ELECTRICITY PRODUCTION COSTS 1995-2008
in 2008 cents per kilowatt-hour

OIL
GAS
COAL
NUCLEAR

Production Costs = Operations & Maintenance + Fuel. Production cost do not include indirect costs or capital.
Source: Ventyx Velocity Suite, via NEI

74103-11_F38.EPS

Figure 38 Electricity production costs.

Act of 2005. The design lays out the best responses to such situations.

1. The terrorist-related events of September 11, 2001

2. Physical, cyber, biochemical, and other potential threats

3. Possible scenarios for attacks by multiple co-ordinated teams directed by a larger group

4. The potential for inside assistance in an attack

5. Suicide attacks

6. Possible water- and air-based threats

7. Possible use of large explosive devices and other modern weapons

8. Possible attacks by persons knowledgeable about facility operations

9. The potential for fires, especially long-burning fires

10. The potential for attacks on spent fuel shipments by multiple coordinated teams

11. Plans to protect public health and safety in a range of nuclear facilities in the event of a terrorist attack

12. The potential for theft of nuclear material from nuclear facilities

All operating nuclear power plants undergo force-on-force (FOF) exercises every three years. The NRC develops appropriate responses to the 12 major threats and conducts drills and mock combat situations between two opposing teams. The defensive team consists of plant security forces with offensive provided by a trained team of outsiders.

The exercises last several days. Plants receive advance notice so two security teams will be present, one for the exercise and one to provide normal security. In addition to plant security, the exercises include federal, state, and local law enforcement personnel and emergency responders. During the mock attacks, the outside force attempts to sabotage the system, to disable safety systems, and to reach the reactor core.

There is no pass/fail for FOF exercises. Their purpose is to evaluate and improve plant security programs and to increase awareness of new threat possibilities.

4.4.0 Permitting and Construction

Before a license is issued, the NRC conducts intensive safety and environmental reviews and also considers the application from an antitrust standpoint. Through the Atomic Energy Act, the local community also has a chance to express its

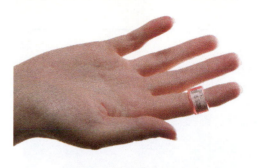

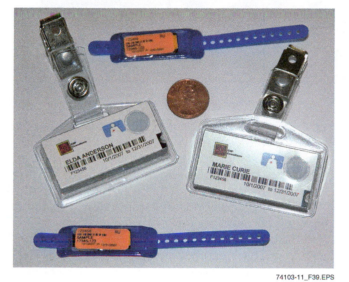

Figure 39 Dosimeters.

74103-11_F39.EPS

opinion. Public hearings must be held, giving people an opportunity to submit written or oral statements to the licensing board.

Protesters who cry "not in my backyard," a movement often referred to as NIMBY, may present their arguments before the board and the public. At the same time, this provides plant and regulatory agency representatives with a means of reassuring the public by outlining plant safety features and explaining security precautions.

Licensing of currently operating commercial nuclear power plants requires both a construction (*Figure 40*) and an operating license. The process is described in Title 10 of the *Code of Federal Regulations (10 CFR)* under *Part 50*. This responsibility falls under the NRC.

74103-11_F40.EPS

Figure 40 Containment building construction, nuclear power plant.

Table 3 Typical Annual Radiation Exposure Sources

Exposure Source	Population Exposed (millions)	Average Dose Equivalent to Exposed Population (millirems/year)	Average Dose Equivalent to U.S. Population (millirems/year)
Natural:			
Radon	230	200	200
Other	230	100	100
Occupational	0.93	230	0.9
Nuclear fuel cycle	—	—	0.05
Consumer products:			
Tobacco	50	—	—
Other	120	5 to 30	5 to 13
Medical:			
Diagnostic X-rays	—	—	39
Nuclear medicine	—	—	14
Approximate total	230	—	360

GOING GREEN

Watts Bar Unit 2

When Unit 2 of TVA's Watts Bar Nuclear Power Plant goes online, it will add 1,180 megawatts of virtually carbon-free power to the Tennessee Valley.

74103-11_SA02.EPS

In 1989, the NRC created a more efficient process by issuing a combined construction and operating license, but it specifies certain conditions for operations in the plant. The newer licensing process (*Figure 41*) is described in *10 CFR Part 52*.

Part 52 also provides for early site permits. Under this permit, permission may be granted for a reactor site that does not specify the reactor type if it is one of several standard, certified, pre-approved plant designs. Design approval is good for 15 years.

The NRC is also responsible for continuing oversight of the facility, including construction and operation, for as long as the plant is active. Oversight is to ensure that the facility meets the NRC's regulations for public and environmental health and safety as well as security.

Physical construction time for a nuclear power plant, from start to finish, is estimated at four to five years, but the actual process, from engineering through hearings, approval, and building, can take

up to 15 years or more. The Tennessee Valley Authority's Watts Bar in Tennessee was the last nuclear generating station in the United States to go online. Construction began in 1966, but because of hearings and various delays, Unit 1 did not go online until 1996. Construction on Unit 2 stopped in 1988, but increasing demand for power fueled a review in 2007 and the project was revived. TVA reports that Unit 2 should go into operation in 2012 or 2013.

5.0.0 PAST, PRESENT, AND FUTURE OF NUCLEAR ENERGY

The history of nuclear energy goes back to 1789 when Martin Klaproth, a German chemist, discovered uranium. He named it after the planet Uranus. Today, more than 200 years later, scientists are working on finding new safer, more efficient ways of using nuclear energy.

5.1.0 Past

A number of scientists contributed to the discovery of nuclear fission, but among the most prominent names are Rutherford and Fermi. Both men won Nobel prizes for their work.

5.1.1 Development of Nuclear Energy

In 1917, British chemist and physicist Ernest Rutherford (*Figure 42*) split an atom, but it was not until 1932 that the first controlled nuclear reaction took

---

On Site

Speeding Up the Licensing Process

The US DOE's Nuclear Power 2010 project is making the licensing process easier so economic risk can be reduced and more companies will be encouraged to invest in new plants.

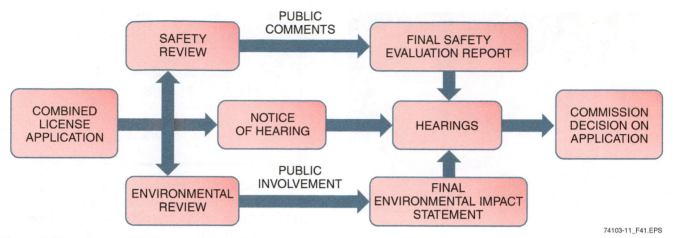

Figure 41 Licensing process.

74103-11_F41.EPS

74103-11_F42.EPS

Figure 42 Ernest Rutherford.

place. Under Rutherford's direction, students John Cockcroft and Ernest Walton split a nucleus in a completely controlled manner.

In the late 1930s, Italian physicist Enrico Fermi discovered that new radioactive elements could be produced by bombarding certain elements with neutrons. In late 1942, Fermi found that inserting cadmium rods, which absorbed neutrons, into the nuclear fission process would slow the reaction. By inserting the rods and then removing some of them, he could generate a self-sustaining reaction and control the rate of fission.

A coded message informed the government of his achievement: "The Italian navigator has just landed in the new world."

5.1.2 Military Influence

When World War II broke out in 1939, it became known that the Germans were working on an atomic bomb. In December of 1942, with the war in the Pacific going badly for the United States, President Franklin Roosevelt authorized a group of scientists to begin work on the same thing. This effort became the Manhattan Project, based in Los Alamos, New Mexico.

Franklin Roosevelt died in office in April 1945, and Harry Truman became president. Following the advice of the military and his cabinet, he made the decision to use atomic weapons to end the war with Japan.

On August 6, 1945, Colonel Paul Tibbets, flying a plane named Enola Gay, dropped an atomic bomb on the Japanese city of Hiroshima. The bomb, code-named Little Boy (*Figure 43*), contained uranium that had been enriched at Oak Ridge, Tennessee.

Three days later on August 9, Major Charles Sweeney flew a B-29 named Bockscar to drop a second bomb, at Nagasaki, Japan. This one, code-named Fat Man (*Figure 44*), contained a plutonium warhead. The plutonium was produced at Hanford Nuclear Site, Washington.

These bombs caused terrifying destruction and a staggering number of deaths. In the wake of the explosions, radiation illness caused many more deaths and health problems.

Those bombings and their terrible images made a lasting impression in the minds of the public. These images are responsible for much of the fear associated with nuclear energy.

Figure 43 Replica of Little Boy.

FAST EXPLOSIVE

NEUTRON INITIATOR

TAMPER/ PUSHER

SLOW EXPLOSIVE

PLUTONIUM CORE

SPHERICAL SHOCKWAVE COMPRESSES CORE

Figure 44 Replica of Fat Man.

5.1.3 Accidents

In almost 60 years since the beginning of civil nuclear power, with well over 400 nuclear reactors in operation worldwide and a cumulative operating experience of more than 14,000 years, only a few

Did You Know?

According to the Nuclear Energy Institute (NEI), nuclear plants hire an average of 1,400 to 1,800 people during the construction period, which lasts several years. During peak times, employment may reach 2,400 construction workers.

major accidents have occurred. The most notable nuclear accidents have occurred at Three Mile Island in March 1979, Chernobyl in April 1986, and Fukushima in 2011.

At Three Mile Island in Pennsylvania, which was built to NRC standards, a very serious situation occurred, but regulation containment facilities and the implementation of proper emergency procedures prevented the escape of radiation; no injuries or deaths occurred.

The Chernobyl Nuclear Power Plant in Ukraine lacked provisions for containment, and the intense fire and release of radiation led to as many as 60 or 70 deaths. The accident resulted from a safety test with inexperienced personnel and a series of errors in judgment. Since the accident, regulations and requirements are much stricter, and there have been no further serious incidents.

In Fukushima, Japan, an earthquake, tsunami, and fires led to the cooling failure of four reactors at the Fukushima 1 Nuclear Power Plant. Within the month following the earthquake, Tokyo Electric Power Company (TEPCO) reported several injuries from the explosion but only minor injuries related to radiation exposure. There were no nuclear-related deaths. The Japan accident has brought new focus to strengthening structures and providing more and better emergency backup systems.

Safety has always been a high priority in nuclear design, and stringent safety procedures are required at all nuclear facilities. In the years since Chernobyl, emphasis on safety has been even greater, particularly regarding public welfare. Safer and more efficient methods of using nuclear power to produce electricity are constantly being developed. These methods are put into practice in the existing industry whenever applicable.

5.2.0 Present

After its initial boom, the nuclear power industry in the United States suffered a series of setbacks and interest waned. Incidents such as Chernobyl, the Cold War threat, the effects of toxic waste (even though there was no connection with nuclear energy), and huge cost overruns in building turned the public against further development.

Now the need for clean, safe energy and energy independence has brought about renewed interest in nuclear energy. As solutions are found to some of the problems, political support has improved and the public is able to consider the pros and cons more objectively.

5.2.1 Career Opportunities

There are many career opportunities in nuclear energy. Concrete and structural steel workers are used heavily during reactor construction, and the demand for new plants is growing. Over the next few years, construction/skilled trade job opportunities in the nuclear energy field are expected to rise considerably faster than in other fields. This is based on the number of new plant permits and planned construction as the demand for more and cleaner energy rises. Current rules concerning operator fatigue add to the need for more plant and reactor operators.

Construction job requirements include a minimum of a high school diploma or GED, background checks, drug tests, relevant work experience, or a technical work certificate. Operator and maintenance technicians are often required to have a two-year degree; however, higher education requirements mean higher salaries. For example, an electrical technician must be able to troubleshoot, test, repair, and maintain sophisticated electrical and electronic equipment as well as typical plant equipment (*Figure 45*). This includes motors, controllers, switchgear, transformers, circuit breakers, generators, and batteries. An electrician is expected to install equipment and control devices.

An instrumentation technician must be able to calibrate, test, troubleshoot, modify, and inspect typical instrumentation found in a nuclear plant. In addition to a high school diploma or GED and two-year degree, employers usually require train-

74103-11_F45.EPS

Figure 45 Cable trays at Bellefont (electrical work).

ing in power plant instrumentation. Many ask for professional certifications.

The military offers excellent training in nuclear energy. Many nuclear careers began there, particularly in the Navy with their nuclear-powered ships and submarines.

Clear career paths offer opportunities for advancement, from electrician to technician to plant operator and reactor operator, with many supervisory positions along the way. Salaries in the nuclear field are generally higher than those for similar positions in non-nuclear fields. Nuclear construction projects may last for years. Building nuclear power plants is a slow process, in part because of the rigorous safety features and requirements. Construction began on the Browns Ferry Nuclear Plant in 1966 (*Figure 46*), but the first unit did not go online until 1973. Currently the plant has three operating boiling water nuclear reactors.

5.2.2 US DOE Position

According to the DOE, their Nuclear Energy (NE) program "promotes secure, competitive, and environmentally responsible nuclear technologies to serve the present and future energy needs of the United States and the world." Both the demand for more energy and current environmental challenges have emphasized the benefits of clean, safe nuclear energy.

One of the DOE's missions is to encourage the nation's nuclear energy research and development program. The focus is on strengthening basic technology and aiding the introduction of the next generation of nuclear power plants.

The Office of Nuclear Energy lists its responsibilities as space and defense nuclear power systems, advanced nuclear research and development, isotope production and distribution, nuclear facilities management, and nuclear fuel security.

On Site

Watts Bar Employment Prediction

As construction on TVA's new Watts Bar unit reaches its peak, approximately 2,300 contract workers will be needed. Once the unit is online, about 290 more permanent jobs will be created at the plant.

Figure 46 Construction begins on TVA's Browns Ferry Nuclear Plant.

The Energy Information Administration (EIA) tracks statistical information on the production and use of nuclear energy.

5.2.3 Regulatory Bodies

Regulatory organizations try to ensure compliance with a wide variety of standards and regulations, and establish the procedures for ensuring and monitoring compliance. For example, the NRC's mission is to license and regulate the nation's civilian use of nuclear materials to ensure adequate protection of public health and safety, to promote the common defense and security, and to protect the environment.

There are many regulatory organizations worldwide, but the predominant ones in the United States include the following:

- Institute of Nuclear Power Operations (INPO)
- Nuclear Energy Institute (NEI)
- Nuclear Regulatory Commission (NRC)

5.2.4 INPO Purpose and Duties

The INPO is a not-for-profit organization established in 1979 following an investigation into the accident at Three Mile Island. The organization is designed to promote the highest levels of safety

On Site

Billions in Loan Guarantees

In February of 2010, President Barack Obama announced more than $8 billion in federal loan guarantees for the construction of the first nuclear power plant in the United States in nearly three decades.

and reliability. INPO accomplishes this through the following activities:

- Establishing objectives, criteria, and guidelines
- Conducting regular detailed evaluations
- Assisting plants in improving performance
- Training and accreditation
- Analyzing events and sharing information

5.2.5 Current Applications

The role of nuclear materials has increased greatly in the last few years. More than one-third of all medical procedures in the United States involve some type of nuclear material; it is used in imaging, treatment, and therapy. Veterinarians use nuclear materials in similar ways, in the treatment of animals.

Research facilities and universities rely more and more on nuclear devices for their work, especially in laboratories.

Industrial uses are growing rapidly. Newer types of gauges, sensors, radiography, all use nuclear materials. *Figure 47* shows an experimental test stand.

Figure 47 Nuclear applications.

5.3.0 Future

In the near future in the Unites States and throughout the world, a huge expansion in the number of nuclear power plants is projected. The main reason is that nuclear power generation is a major source of carbon-free energy. By the year 2020, fossil-fuel-based electricity will be responsible for more than 40 percent of greenhouse gas emissions worldwide.

To quote NRC Commissioner William C. Ostendorff, on October 4, 2010:

> The NRC has received 18 license applications for 28 new nuclear power plants (*Figure 48*); of these, 13 applications for 22 units are under active NRC review. Five applications have been suspended or deferred by the applicants because of their changing business strategies. One of the five suspended applications has since been withdrawn and replaced with an application for an early site permit. These are the first applications for new reactors that the NRC has received in roughly three decades.

On Site

Nuclear Fleet Buildup

For an in-depth report on nuclear power, see "The Future of Nuclear Power: The United States. and the world is gearing up to build a potentially massive fleet of new nuclear reactors, in part to fight climate change. But can nuclear power handle the load?" *Scientific American.* January 26, 2009.

However, problems remain that will hinder development until solutions are found. The public is still skeptical about the effects of nuclear power on health, the environment, and safety. There are risks with nuclear proliferation; security is a major factor. The relative capital costs of building a nuclear power plant discourage many commercial investors. The last obstacle is how to manage nuclear wastes, but research into new reactors that will be more efficient and produce less waste is promising.

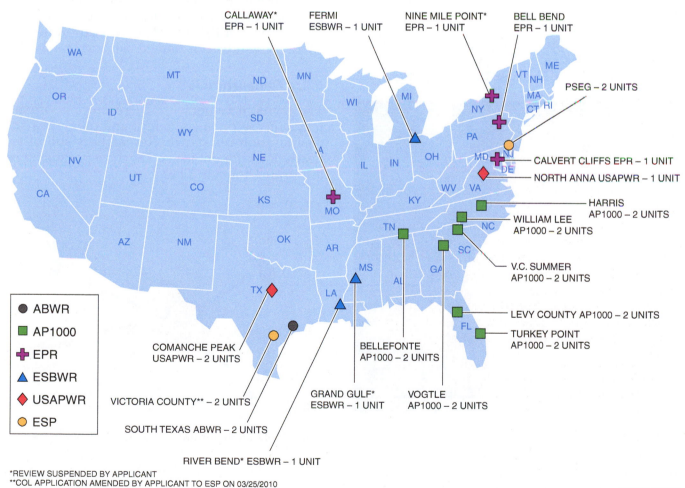

Figure 48 New reactor locations.

*REVIEW SUSPENDED BY APPLICANT
**COL APPLICATION AMENDED BY APPLICANT TO ESP ON 03/25/2010

74103-11_F48.EPS

5.3.1 Fusion

Fusion is the process in which atomic nuclei move so fast they collide and stick together, emitting a huge amount of energy. When nuclei are exposed to extremely high temperatures, the ions move faster and faster until they reach speeds that bring them close enough together that the nuclei fuse. When the nuclei fuse, an enormous amount of energy is released in the form of heat. When scientists and engineers perfect this process, the benefits will be tremendous in the amount of clean energy generated.

The sun and stars are powered by fusion. In that process, hydrogen atoms fuse together and form helium, converting matter into energy. At very high temperatures, the hydrogen changes from a gas to plasma and positive and negative electrons are separated from each other. In *Figure 49*, the hottest areas appear almost white, while the darker red areas indicate cooler temperatures. The swirls on the right are huge clouds of dense plasma suspended in the sun's hot, thin corona. The flares and loops appear darker because, compared to the inner temperatures, they are cooler.

In most applications, fusion is not yet possible. One reason is that the repulsive forces in positively charged nuclei are too strong to allow them to get close enough to fuse. A simple example of this is trying to force the positive ends of two magnets together. The resistance is strong, and the more powerful the magnets, the harder it is to make them touch. The second reason is that the temperatures needed for nuclear fusion are far too high and would destroy any material container.

Two types of experimental reactors for nuclear fusion are being considered. One is called magnetic confinement and the other is inertial confinement. In magnetic confinement, the hot plasma is contained by strong magnetic fields. In inertial confinement, strong lasers or particle beams compress small pellets made of fusion fuel into extremely dense material. During compression, fusion occurs, giving off huge amounts of energy.

At present most research is based on a mix of deuterium and tritium; both are heavy isotopes of hydrogen. *Figure 50* shows a simplified illustration of the deuterium-tritium-hydrogen fusion process.

5.3.2 Research and Development

By 2050, Earth is expected to house about 10 billion people. The demand for energy will grow as the population increases. To meet these demands, ten countries, including the Republic of South Africa, Switzerland, the United Kingdom, and the United States, agreed to work together in an international effort to develop clean, safe, cost-effective energy. A large part of this effort is focused on nuclear energy. The project is named Generation IV.

5.3.3 Generation IV

The focus of Generation IV reactors is on sustainability, economics, safety, and reliability, and proliferation resistance. Researchers submitted nearly 100 different ideas. The most promising reactors are shown in *Table 4*.

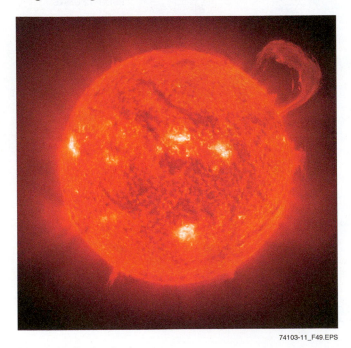

74103-11_F49.EPS

Figure 49 Fusion in the sun.

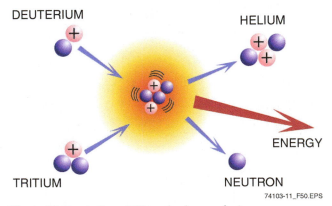

74103-11_F50.EPS

Figure 50 Deuterium-tritium-hydrogen fusion process.

Table 4 IV Reactor Types

Reactor Type	Fuel	Expected Deployment Date
Gas-cooled fast reactor (GFR) system	Fast-neutron spectrum and closed fuel cycle for efficient conversion of fertile uranium	2025
Molten salt reactor (MSR) system	Thermal neutron spectrum and closed fuel cycle for efficient use of plutonium	2025
Sodium-cooled fast reactor (SFR) system	Fast-neutron spectrum and a closed fuel cycle for efficient conversion of fertile uranium	2015
Supercritical water-cooled reactor (SCWR) system	Two fuel cycle options: • Open cycle with thermal neutron spectrum reactor • Closed cycle with a fast-neutron spectrum reactor and full actinide recycle Both use high-temperature and pressure, water-cooled reactor that operates above the thermodynamic critical point of water	2025
Very high temperature reactor (VHTR) system	Thermal neutron spectrum and a once-through uranium cycle	2020

SUMMARY

Nuclear power is a source of energy that can produce massive amounts of emission-free electricity. Nuclear energy is produced when an atom is split. The splitting of an atom, or fission, is the source of nuclear energy. Only some elements have atoms that can be split. Uranium is the element most easily split. When an atom splits, it releases a tremendous amount of energy, usually in the form of heat. In a power plant, this heat is used to make steam. The steam is used to generate electricity.

Nuclear power plants have special structures and equipment where fission takes place, so that it can be controlled and protected. Safety is always the primary concern. The public and the environment are carefully protected. Atoms are split inside strong, sealed structures called reactors. There are several types of reactors, but they all produce steam that is converted to mechanical energy, which generates electricity.

Among the advantages of nuclear power are the low life-cycle cost, available technology, lack of harmful carbon emissions, and very small amount of fuel needed to produce a great deal of power.

Its disadvantages include storage and disposal of hazardous wastes. Rigid safety regulations drive the cost of building nuclear plants high, but once they are built, the cost of producing electricity is usually lower than that of fossil-fuel plants.

1. In 2009 in the United States, approximately 21 percent of the electricity was generated by _____.

   a. coal
   b. gas
   c. nuclear energy
   d. hydroelectric power

2. Uranium is the heaviest naturally occurring element, having an atomic number of _____.

   a. 92
   b. 146
   c. 235
   d. 238

3. When the number of neutrons does *not* match the number of protons, the element is called a(n) _____.

   a. isotope
   b. nucleus
   c. stable element
   d. subatomic particle

4. In the milling process, the waste rock is called _____.

   a. yellowcake
   b. tailings
   c. leach product
   d. uranium ore

5. The cruciform-shaped devices that are inserted and removed from the core to control the rate of fission are the _____.

   a. moderators
   b. control rod assemblies
   c. reactor cores
   d. fuel assemblies

6. The concrete structure several feet thick that surrounds the pressure vessel and reactor core is called the _____.

   a. containment vessel
   b. safety structure
   c. reactor housing
   d. steam generator

7. In a PWR, the water in the reactor is kept under pressure to maintain the heat and prevent it from _____.

   a. boiling
   b. circulating
   c. absorbing neutrons
   d. becoming radioactive

8. When water is heated, the molecules become more active and _____.

   a. absorb more heat
   b. become radioactive
   c. draw closer together
   d. expand and take up more space

9. The largest source of nuclear pollutants comes from _____.

   a. spent fuel rods
   b. contaminated water
   c. uranium mine tailings
   d. contaminated tools and clothing

10. The time it takes for half the atoms in a radioactive element to break down is its _____.

    a. half-life
    b. waste potential
    c. contamination ratio
    d. useable nuclear time

11. In 30 years of transporting spent fuels in the United States, the number of radioactive releases that have occurred is _____.

    a. none
    b. two
    c. three
    d. four

12. In 1917, British chemist and physicist Ernest Rutherford split a(n) _____.

    a. atom
    b. isotope
    c. neutrino
    d. molecule

13. In the late 1930s, Italian physicist Enrico Fermi discovered that new radioactive elements could be produced by bombarding certain elements with _____.

    a. atoms
    b. protons
    c. neutrons
    d. molecules

14. In 1945, three days after the atom bomb was dropped on Hiroshima, a second bomb was dropped at Nagasaki, Japan, containing a _____.

    a. uranium warhead
    b. plutonium warhead
    c. deuterium warhead
    d. thorium warhead

15. The capacity of nuclear power generation is expected to increase when scientists perfect the process of combining atomic nuclei by _____.

    a. fission
    b. radioactivity
    c. fusion
    d. compression

Name: _____

Date: _____

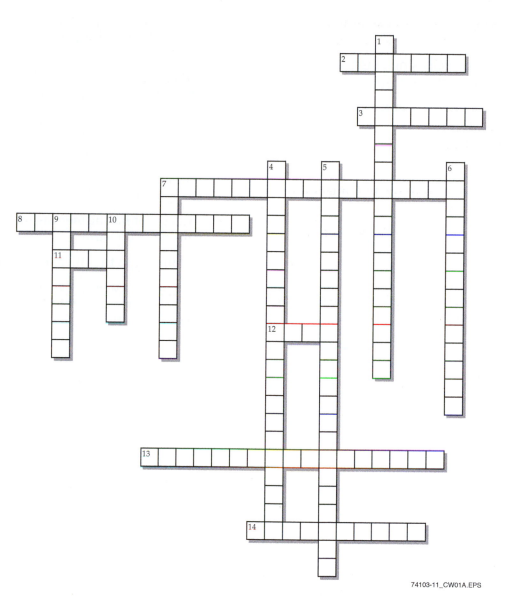

74103-11_CW01A.EPS

Across:

2. An individual particle of electromagnetic energy

3. Splitting atoms, releasing massive amounts of energy

7. A reactor that uses water laced with deuterium to generate energy

8. Conversion of a material into hard glass

11. A measurement of radiation dosage to the human body related to the potential damage

12. The area of a reactor where fission takes place

13. Uses plain water in its processes

14. A mining process to break rock, usually by high pressure fluid injection

Down:

1. A light water type that separates the steam rising off of boiling water to drive a turbine

4. An upright structure for cooling water, shaped with a "waist line"

5. A light water reactor that keeps the water under enormous pressure, allowing it to become superheated

6. The kind that creates more fissionable nuclear material than it uses

7. Water that has numerous molecules containing deuterium

9. Belonging to the third level or order

10. The process where the nucleus of atoms collide with sufficient force to cause them to stick together

Cornerstone of Craftsmanship

Tony Vazquez

President/CEO VisionQuest-Academy, Ocala, FL

Tony Vazquez came from a family of furniture-makers that emigrated from Cuba to the United States. Influenced by his father and grandfather, he entered the construction industry. It wasn't long before he and his father opened their own company, but the need to find good craftsmen led him to training. As the need for quality training became more apparent, he eventually opened his own construction training school. Tony enjoys sharing his love of building and encouraging others to excel in the industry that's made him a success.

How did you get started in the construction industry and who inspired you?

I was raised in a second-generation, construction/ craft, working family. I was inspired as a boy by my father and grandfather back in Cuba, where they hand-made furniture. Later, here in the United States, my father continued his craft as a construction worker and I, as a teenager, often went with him on weekends. After I left high school, my father and I started a construction company. The hard work that came with the job was outweighed by the gratification of seeing a well-built project that would last for many years. In 1990, while looking for trained construction workers, I ended up at the Miami Job Corps. I was impressed with the training going on in our industry. Later that year I volunteered my time to help with their program, and a few years later when the instructor retired, I was asked to take over. I returned to college and obtained my instructor/teacher certification in the industry and have enjoyed training others ever since.

What do you enjoy most about your job?

I still love to build. However, today I get the most satisfaction from teaching others the art of craftsmanship. I have taught at both the secondary (high school) and post-secondary (adult) levels for years. For decades the basics of construction have been taught at a few high schools (shop class) and a few post-secondary training programs scattered throughout the country. As our industry becomes more united in providing an industry-driven curriculum and nationally recognized training, it is revitalizing the

art of construction craftsmanship. It has come during a much-needed time as our older workforce is preparing to retire and we have need of new blood and craftsmanship in our industry.

Do you think training and education are important in construction?

Yes, training and education are at a most influential time. Tough times today require specialized training. In today's way of building, each trade has become specialized, where those with the most training, experience, and credentials get the job. As our industry moves forward in creating nationally recognized credentials and certifications, so too have specialized training centers and schools at both secondary and post-secondary levels.

How important are NCCER credentials to your career?

Extremely important, in two distinct and different ways. First, by establishing standard trainer credentials, the industry ensures that training taking place across the country is similar and of the same quality. Second, completion points and credentials given to trainees are also similar and of the same quality across the country, thus making it possible to have national certification that is valid anywhere. At a personal level, having and acquiring multiple credentials allows me to excel. Reaching Master Trainer and Subject Matter Expert status with NCCER in the construction and green job trade has given me the opportunity to create my own construction training academy.

How has training/construction impacted your life and your career?

Construction training has impacted my life since my teen years, both informally and formally. In the early years it provided the technical background for the work I was performing. Years later, training became more formal in specialized areas. Currently, the combination of field work/training and formal classroom training has provided me with a good salary and lifestyle. Furthermore, it has provided me the opportunity to open VisionQuest-Academy, a full-service, post-secondary educational/training facility in Ocala, Florida. The academy is dedicated to providing cutting-edge construction and green job training, and courses are offered in English and Spanish. NCCER has been a key to this opportunity; with its support, many of our county's adults/working population now have the opportunity obtain national industry credentials.

Would you suggest construction as a career to others?

Yes, absolutely. As I tell my students, both youth and adults, it's never too late for training and education, either informal or formal. However, a well-planned training strategy is like a well-planned trip that includes a road map to your destination. NCCER has that covered for the construction and green job industry along with pathways to success. Ultimately, the construction and green job industry is especially rewarding in that, at the end of the day, you have something tangible to look back on.

How do you define craftsmanship?

Craftsmanship can be defined in a multitude of ways. For the most part, in the construction industry, craftsmanship is the final product of a job, task, or project completed within the quality standards set for that industry. However, I define it as follows: the skills, expertise, ability, and technique used by a person to shape, mold, transform, and convey an idea into products of value to others. A craftsman's level of work and performance can only be achieved by a combination of hard work, informal and formal training, time, and mostly a love of the craft.

Trade Terms Introduced in This Module

Boiling water reactor (BWR): A light water nuclear reactor in which the water boils in the reactor core and is drawn up through a separator in the top of the pressure vessel. The water is separated from the vapor, which is sent directly to a steam turbine.

Breeder reactor: A reactor that creates more fissionable material than it uses.

Core: The central part of a nuclear reactor where fission occurs.

Fission: The splitting of atoms, which releases a tremendous amount of energy in the form of heat.

Fracturing: A mining process used to break or separate rock, usually by injecting fluid into a hole, either natural or manmade, under high pressure.

Fusion: The process in which atomic nuclei collide so fast they stick together and emit a large amount of energy. In the center of most stars, hydrogen fuses into helium.

Heavy water: Water laced with a large number of molecules that contain deuterium atoms; deuterium is also called heavy hydrogen.

Heavy water reactor: In most reactors, the moderator is plain water; but some reactors use heavy water, which does not absorb neutrons as regular hydrogen does, making it useful in slowing neutrons from fission.

Light water reactor: A nuclear reactor that uses plain water as a coolant and moderator.

Parabolic cooling tower: An upright cooling tower with a "waist." The shape induces a natural draft, drawing air in at the bottom and forcing it out at the top.

Photons: An individual particle of electromagnetic energy such as light; a basic unit that has no mass.

Pressurized water reactor (PWR): A type of light water reactor is which the water is kept under pressure and not allowed to expand, so it gets hotter and hotter but stays in its liquid state, creating enormous pressure. The superheated, pressurized water is carried to a steam generator where it is used to produce steam.

Roentgens per man (rems): A quantity that relates to the amount of damage to the human body an absorbed dose of radiation could (but does not necessarily) cause. It is often expressed in thousandths of a rem, or millirems.

Tertiary: Belonging to the third level or order.

Vitrification: The conversion of a material into extremely hard glass by subjecting it to temperatures lower than its normal melting point.

Additional Resources

This module presents thorough resources for task training. The following resource material is suggested for further study.

US Energy Information Administration. www.eia.gov

Center for Energy Workforce Development. www.getintoenergy.com

US Department of Energy. www.energy.gov

US Department of Energy, Office of Nuclear Energy. www.ne.doe.gov

Institute of Nuclear Power Operations. www.inpo.info

Nuclear Energy Institute. www.nei.org

US Nuclear Regulatory Commission. www.nrc.gov

Coolclean Cooling Towers. www.coolclean.com.au

Duke Energy. www.duke-energy.com

Figure Credits

Cameco Corporation, Module opener and Figure 11

Topaz Publications, Inc., Figures 1 and 17

Courtesy of US Department of Energy, Figures 5, 29, 30, 32, 33, 37, and AIG Project 3

Photo by Andrew Silver, US Geological Survey, Figure 6

Nuclear Energy Institute, Figures 7, 10, 15, and 40

NRC, Figures 8, 9, 12–14, 20–22, 24, 34-36, 41, 45, 48, SA02, Table 4, and AIG Projects 2 and 4

Xcel Energy, Figure 18

Dresser-Rand, Figure 25

John J. Mosesso/life.nbii.gov, Figure 27

Canadian Nuclear Association, Figure 31

Ventyx, an ABB Company and Nuclear Energy Institute (NEI), Figure 38

Mirion Technologies, Figure 39 (top photo)

CHP Dosimetry, Figure 39 (bottom photo)

US Air Force photo, Figures 43 and 44

Tennessee Valley Authority, Figure 46

US Air Force photo/Lance Cheung, Figure 47

SOHO (ESA & NASA), Figure 49

NCCER CURRICULA — USER UPDATE

NCCER makes every effort to keep its textbooks up-to-date and free of technical errors. We appreciate your help in this process. If you find an error, a typographical mistake, or an inaccuracy in NCCER's curricula, please fill out this form (or a photocopy), or complete the online form at **www.nccer.org/olf**. Be sure to include the exact module ID number, page number, a detailed description, and your recommended correction. Your input will be brought to the attention of the Authoring Team. Thank you for your assistance.

Instructors – If you have an idea for improving this textbook, or have found that additional materials were necessary to teach this module effectively, please let us know so that we may present your suggestions to the Authoring Team.

NCCER Product Development and Revision
13614 Progress Blvd., Alachua, FL 32615

Email: curriculum@nccer.org
Online: www.nccer.org/olf

❏ Trainee Guide   ❏ AIG   ❏ Exam   ❏ PowerPoints   Other _____

Craft / Level: _____   Copyright Date: _____

Module ID Number / Title: _____

Section Number(s): _____

Description: _____

_____

_____

_____

Recommended Correction: _____

_____

_____

_____

Your Name: _____

Address: _____

Email: _____   Phone: _____

74104-11

Solar Power

Module Four

Trainees with successful module completions may be eligible for credentialing through NCCER's National Registry. To learn more, go to **www.nccer.org** or contact us at **1.888.622.3720**. Our website has information on the latest product releases and training, as well as online versions of our *Cornerstone* newsletter and Pearson's product catalog.

Your feedback is welcome. You may email your comments to **curriculum@nccer.org,** send general comments and inquiries to **info@nccer.org**, or use the User Update form at the back of this module.

Copyright © 2011 by the National Center for Construction Education and Research (NCCER) and published by Pearson Education, Inc., publishing as Prentice Hall. All rights reserved. Manufactured in the United States of America. This publication is protected by Copyright, and permission should be obtained from NCCER prior to any prohibited reproduction, storage in a retrieval system, or transmission in any form or by any means, electronic, mechanical, photocopying, recording, or likewise. To obtain permission(s) to use material from this work, please submit a written request to NCCER Product Development, 13614 Progress Blvd., Alachua, FL 32615.

Objectives

When you have completed this module, you will be able to do the following:

1. Define solar power, how it is harnessed, and how it is used to generate energy.
2. List the advantages and disadvantages of solar energy.
3. Describe the past, present, and future of solar energy.
4. Identify and describe solar applications.

Performance Tasks

This is a knowledge-based module; there are no performance tasks.

Trade Terms

Air mass
Altitude
Ambient temperature
Amorphous
Array
Autonomy
Azimuth
Backfeed
Balance of system (BOS)
Brownout
Building-integrated photovoltaics (BIPV)
Charge controller
Combiner box
Concentrating collector
Declination
Depth of discharge (DOD)
Doped
Dual-axis tracking
Electrochemical solar cell
Elevation
Evacuated tube collector
Flat plate collector
Fuel cells
Grid-connected system
Grid-interactive system
Grid-tied system
Heliostats
Hybrid system

Insolation
Integral-collector storage system
Inverter
Irradiance
Latitude
Maximum power point tracking (MPPT)
Module
Monocrystalline
Net metering
Off-grid system
Peak sun hours
Polycrystalline
Pulse width-modulated (PWM)
Sea level
Semiconductor
Sine wave
Single-axis tracking
Solar photovoltaic (PV) system
Solar thermal system
Spectral distribution
Standalone system
Standard Test Conditions (STC)
Sun path
Thermosiphon system
Thin film
Tilt angle
Utility-scale solar generating system
Watt-hours (Wh)

Industry Recognized Credentials

If you're training through an NCCER-accredited sponsor you may be eligible for credentials from NCCER's Registry. The ID number for this module is 74104-11. Note that this module may have been used in other NCCER curricula and may apply to other level completions. Contact NCCER's Registry at 888.622.3720 or go to nccer.org for more information.

Contents ─────────────────────────

Topics to be presented in this module include:

Figures and Tables

1.0.0 INTRODUCTION

Solar power harnesses the energy of the sun and uses it to produce heat or electricity. Solar thermal systems use sunlight to heat fluids or spaces. Solar photovoltaic (PV) systems convert sunlight into electricity. When sunlight is used directly to produce heat, it is known as passive solar power. When sunlight is used to produce another form of energy, such as electricity, it is known as active solar power. While active forms of solar power have only become common since the mid-1950s, passive solar power has been in use for over two thousand years.

1.1.0 History of Solar Power

The ancient Greeks used thick walls to trap heat during the day and release it slowly at night. They also oriented buildings to provide shading in summer while maximizing sunlight in winter. Solar energy was also used to preserve food, heat water, and dry clothes. These early applications of solar power were very effective and are still in use today.

In 1767, Horace de Saussure experimented with glass boxes to determine how much heat could be trapped by the glass. He discovered that heat could be collected on sunny days even if the outdoor temperature was quite low, as on a mountaintop. These experiments helped scientists understand the effects of atmospheric differences on outdoor temperature and also became the basis for passive solar collectors and solar ovens.

Active forms of solar power did not come into use until 1839, when Alexander Becquerel discovered that certain materials produce electric current when exposed to light. This eventually led to the invention of the first photovoltaic (PV) cell.

In 1891, Clarence Kemp patented the first water heater powered by solar energy. Within a few years, many homes in California had hot water supplied by the sun. In 1909, William Bailey patented a solar water heating system with a separate storage tank that allowed the water to be stored in larger quantities. Solar water heaters remained popular until the discovery of natural gas in California in the 1920s. Like other forms of alternative energy, the popularity of solar power is tied to the availability of other fuel sources. With the discovery of inexpensive natural gas, the use of solar thermal power declined in California. It remained popular in other sunny states such as Florida, but fell out of use when the price of electricity dropped after World War II.

The first practical PV cell was invented by Bell Laboratories in 1954 using modified silicon cells. A PV cell consists of two layers of semiconductor, one p-type (positive) and the other n-type (negative). The electrical contacts are screened in a grid on both sides of the panel. High-energy light particles known as photons strike the semiconductor atoms and release free electrons, producing current (*Figure 1*). Many cells are combined in a solar panel or module, and these modular panels are connected in an array.

As America entered the space race in the 1950s, scientists realized that solar arrays provided an ideal power source for satellites. Later, solar cells

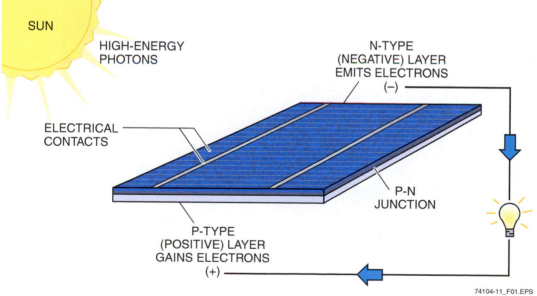

SUN

HIGH-ENERGY PHOTONS

N-TYPE (NEGATIVE) LAYER EMITS ELECTRONS (–)

ELECTRICAL CONTACTS

P-N JUNCTION

P-TYPE (POSITIVE) LAYER GAINS ELECTRONS (+)

74104-11_F01.EPS

Figure 1 Photovoltaic process.

Did You Know?

Solar energy is still the primary source of power in satellite systems and spacecraft. This photo shows the array on a spacecraft designed to map the location of our solar system in the Milky Way galaxy.

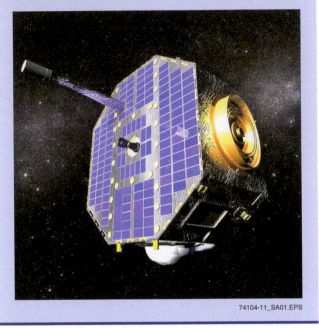

74104-11_SA01.EPS

became popular for operating rural telephones and radio transmitters, and for homes that were too remote to be served by the electrical grid.

In the 1970s, the energy crisis focused more attention on alternative energy sources, and the use of solar power increased in both residential and commercial applications (*Figure 2*). Today, active solar power is used for everything from handheld calculators to giant solar thermal power plants (*Figure 3*).

1.2.0 Advantages and Disadvantages

Solar power has many advantages over other energy sources. Sunlight is a limitless resource, and most areas can support some degree of solar power. Solar power is clean and environmentally friendly. It can be harnessed without disrupting the environment and produces no hazardous waste or emissions. It is also quiet and requires little maintenance. In addition, solar power can be generated on site, and does not have the mining/drilling or transportation requirements of fossil fuels. The power generated is also a domestic product, which strengthens the economy, produces jobs, and reduces reliance on foreign energy.

Solar PV systems help to offset power use. Putting energy back into the grid reduces the possibility of **brownouts**. As electricity is produced, it is used at the home or business first, and then any excess electricity is sent into the grid. This is known as **net metering**. Customers receive credits for generating power, which can help to offset

74104-11_F02.EPS

Figure 2 Typical commercial PV system.

Figure 3 Solar thermal power plant.

the initial cost of installing a PV system. In addition, both PV and solar thermal customers may receive government incentives and utility rebates when installing these systems.

The main disadvantages of solar PV/thermal systems include the high initial cost, space requirements, and the need for exposure to sunlight.

Solar panels require significant space, either on a rooftop or on the ground. If batteries are used with a solar PV system, they must be stored in a protected area. Some types of batteries are more hazardous than others. As the technology develops, more power will be produced using fewer panels, reducing both the cost and space required for installation.

Many people believe that solar PV panels can only be installed in hot, sunny locations, but this is not the case. PV panels actually produce more voltage at cooler temperatures, and there have been many advances in their use in cloudier climates. However, the energy produced is limited by the available sunlight, both throughout the day and over the seasons. For year-round use, this requires some form of backup power. With a solar thermal system, a backup water heater is required.

1.3.0 Career Opportunities

The use of solar power is now increasing by more than 25 percent per year, creating a huge demand for certified installers. Other career opportunities include estimating, maintenance and troubleshooting, and manufacturing.

Think About It

Find the cost of an average solar system in your area.

Did You Know?

Earth receives more energy from the sun in a single hour than the world uses in a whole year.

Source: Massachusetts Institute of Technology

PV and solar thermal system installation often involves working in extreme heat on slanted, slippery roof surfaces ten or more feet above the ground (*Figure 4*). In addition to working at heights, PV installers also handle sharp, unwieldy panels that become energized the moment light strikes them (*Figure 5*). Installers must follow all safety precautions and be in good physical condition for this type of work. In addition, estimators must have above-average mathematical skills to perform the load calculations. Electrical skills are required for making system connections. Plumbing skills are required when working on solar thermal systems.

74104-11_F04.EPS

Figure 4 PV array installed on a terra cotta roof.

74101-11_F05.EPS

Figure 5 Installing PV panels.

2.0.0 SOLAR THERMAL APPLICATIONS

Have you ever been surprised by how hot the water becomes when a garden hose is left lying in the sun? This is the basic principle of a solar thermal system at work. In these systems, the sun's energy is used to heat water, air, or another fluid. Solar thermal systems can be used in low-temperature, medium-temperature, and high-temperature applications.

2.1.0 Low-Temperature Applications

Low-temperature systems heat water or air to a temperature of less than 110°F (43°C). They are used for applications such as space heating or as solar pool heaters. They are also used on a limited basis for heating water.

2.2.0 Medium-Temperature Applications

Medium-temperature systems heat water or air to a temperature between 110°F (43°C) and 180°F (82°C). They are often used to heat water for home use. In these systems, water is pumped through flat panels exposed to the sun. The water is then returned to a holding tank for use. Solar thermal systems can be used to provide all of a home's hot water during the summer months and a good portion of it throughout the year, depending on the local climate. In larger systems, the heated water can also be supplied to radiators to reduce the heating load.

Solar water heating systems include storage tanks and solar collectors. Solar collectors gather the sun's energy, transform it into heat, and then transfer that heat to the water in the collector. Two basic types of solar collectors are used in solar water heating systems: **flat plate collectors** and **evacuated tube collectors**.

A flat plate collector (*Figure 6*) is a metal box with a transparent cover that contains three main components: an absorber to collect the heat, tubes containing water or air, and insulation. The water or air is warmed as it passes through tubes below the absorber plate and returned indoors using a pump or fan.

Evacuated tube collectors consist of parallel rows of transparent glass tubes (*Figure 7*). Each tube contains a glass outer tube and metal inner tube that acts as the absorber. A vacuum between the tubes preserves heat, similar to a glass thermos bottle. Evacuated tubes are more efficient than flat plate collectors and also work better under cloudy conditions. For this reason, they are often used in colder climates. They are also more

Zero Energy Homes

Some homes use a combination of active and passive solar energy, along with weatherization and other techniques, to produce as much energy as they use. They are known as zero energy homes, one of which is shown here. This home is in Denver, Colorado. The roof overhangs are designed to shade the windows from the sun in the summer, but allow sunlight to provide light and warmth in the winter when the angle of the sun is lower. Over the course of a year, the energy produced by the solar array equals the energy required to operate the home, resulting in a net use of zero.

74104-11_SA02.EPS

expensive and fragile, since the tubes are not protected by an outer case.

Solar water heating systems require a backup system for use in poor weather and when the system cannot produce enough hot water to meet the demand. They also require a storage tank. In two-tank systems, the solar water heater preheats water before it enters the regular water heater. One-tank systems combine the backup heater and the solar storage in one tank.

There are two types of solar water heating systems: active, which use pumps to circulate the water, and passive, which rely on gravity. Both types are shown in *Figure 8*.

74104-11_F06.EPS

Figure 6 Flat plate collector.

74104-11_F07.EPS

Figure 7 Evacuated tube collector.

2.2.1 Passive Solar Water Heating Systems

The simplest type of solar thermal system is a passive system. Passive solar water heating systems are less expensive than active systems, but they are also less efficient. Passive solar water heaters rely on gravity and the natural circulation of heated water to operate. There are two basic types of passive systems: **integral-collector storage systems** and **thermosiphon systems**.

Integral-collector storage systems consist of one or more storage tanks in an insulated box with a glazed side facing the sun. These systems work best in areas where temperatures rarely fall below freezing. They also work well in homes where the hot water needs, such as for laundry and bathing, are highest in the evening rather than the morning. This is because they lose most of their stored energy overnight.

Thermosiphon systems rely on convection to circulate water through the collectors and to the tank (*Figure 9*). As water in the solar collector heats, it becomes lighter and rises into the tank. This pushes the cooler water down to the bottom of the collector. Some thermosiphon systems use glycol to prevent the water from freezing in colder climates.

Did You Know?

A homeowner using an electric water heater could save up to $500 per year by installing a solar water heating system. The use of solar power to heat water is so valuable that some countries have made its use mandatory. For example, all homes in Israel have solar hot water systems.

Source: www.consumerenergycenter.org

2.2.2 Active Solar Water Heating Systems

There are two types of active solar water heating systems, including direct circulation and indirect circulation. Direct-circulation systems use pumps to circulate pressurized water directly through the collectors. These systems are used in areas that do not freeze for long periods and do not have hard or acidic water.

Indirect-circulation systems pump heat-transfer fluids through collectors. Heat exchangers transfer the heat from the fluid to the water. Since the heat transfer fluid is resistant to freezing, these systems are popular in colder climates.

2.3.0 High-Temperature Applications

High-temperature solar thermal systems heat water or other fluids to a temperature of 180°F (82°C) or higher. They are used in utility-scale solar thermal generating systems. These systems are ideal for installation in sunny, uninhabited areas such as deserts. The solar power plants in California's Mojave Desert are the world's largest solar power plants in operation today.

Utility-scale solar thermal systems include parabolic troughs, parabolic dish systems, and central power towers. The word *parabolic* simply means curved toward a center point, like a satellite dish. These systems use a special type of collector known as a **concentrating collector**, which is a system of lenses or mirrors used to focus solar energy on a receiver.

In parabolic trough applications, each trough is equipped with its own receiver that can focus sunlight at 30 to 100 times its normal intensity (*Figure 10*). The heat is transferred to a central location, where it produces steam. The steam is then used to drive a turbine and generate electricity, similar to a fossil fuel generator except without the dangerous emissions or depletion of natural resources. This type of system is known as a solar boiler.

In parabolic dish systems (*Figure 11*), the sunlight can be focused to more than 2,000 times its normal intensity with temperatures over 1,380°F (749°C). The engine in a solar dish/engine system converts heat to mechanical power using a turbine or piston. The engine is attached to an electric generator to convert the mechanical power to electrical power. The advantage of a dish reflector is that the engine is part of the system, which allows for independent power production in remote areas. This is also a disadvantage, as it requires a heavy support structure.

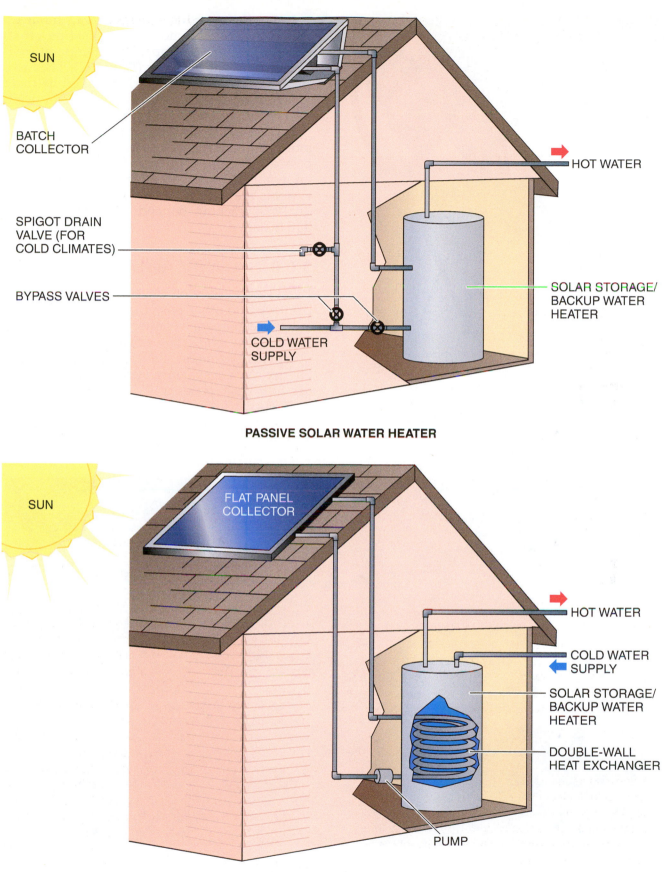

SUN

BATCH
COLLECTOR

SPIGOT DRAIN
VALVE (FOR
COLD CLIMATES)

BYPASS VALVES

COLD WATER
SUPPLY

HOT WATER

SOLAR STORAGE/
BACKUP WATER
HEATER

PASSIVE SOLAR WATER HEATER

SUN

FLAT PANEL
COLLECTOR

HOT WATER

COLD WATER
SUPPLY

SOLAR STORAGE/
BACKUP WATER
HEATER

DOUBLE-WALL
HEAT EXCHANGER

PUMP

ACTIVE SOLAR WATER HEATER

74104-11_F08.EPS

Figure 8 Passive and active solar thermal systems.

Power tower systems operate in a similar way, with an intensity of more than 1,500 times that of normal sunlight and operating temperatures of 600°F (316°C) and higher (*Figure 12*). They use a system of sun-tracking mirrors called **heliostats** to focus the sunlight on a central receiver used to heat water or liquid sodium to drive a turbine.

Parabolic dish and power tower systems use a motor to track the sun in both height and compass direction as it travels across the sky. This is known as **dual-axis tracking**. Parabolic trough systems may use either **single-axis tracking** or dual-axis tracking, depending on the type of system installed.

3.0.0 SOLAR PV APPLICATIONS

PV systems are often classified by how they are connected to other power sources and loads. PV systems can operate connected to or independent of the utility grid. They can also be connected to other energy sources, such as wind turbines, and energy storage systems, such as batteries. There are four basic types of PV systems: **standalone systems**, **grid-connected systems**, **grid-interactive systems**, and **utility-scale solar generating systems**.

74104-11_F09.EPS

Figure 9 Thermosiphon system.

74104-11_F10.EPS

Figure 10 Parabolic trough system.

Figure 11 Parabolic dish system.

3.1.0 Standalone Systems

Standalone PV systems can be either direct-drive or battery-powered. These systems are commonly used to provide power in areas where access to the grid is inconvenient or unavailable.

Direct-drive systems power direct current (DC) loads, such as ventilation fans, irrigation pumps, and remote cattle watering systems. Because these systems operate only when the sun is shining, it is essential to match the power output of the array to the load.

Battery-powered standalone systems use PV energy to charge one or more batteries. The battery system then supplies either a DC load or an alternating current (AC) load when an inverter is used to convert the DC to AC. Battery-powered standalone systems range from handheld electronics to trailer-mounted systems used to provide emergency power (*Figure 13*). They can also be used for remote data monitoring, emergency highway signage, and street or parking lot lighting (*Figure 14*).

Standalone systems are also used to power buildings in remote areas. These systems do not

Figure 13 Portable PV generator.

Figure 12 Power tower system.

Figure 14 Solar-powered parking lot fixture.

3.2.0 Grid-Connected Systems

Grid-connected systems operate in parallel with the utility grid. Also known as grid-tied systems, these systems are designed to provide supplemental power to the building or residence (*Figure 15*). Since they are tied to the utility, they only operate when grid power is available.

Grid-tied systems invert the DC produced by a solar array into AC, which is then sent to the building's electrical panel to supply power. A DC disconnect is required between the array and the inverter, while an AC disconnect is required between the inverter and the building service panel.

Grid-tied systems require the installation of a special meter by the utility. During the daytime, any power in excess of the load is sold back to the utility in the form of credits. This can be shown by the building's electrical meter, which effectively runs backward when excess energy is supplied. At night and during periods when the load exceeds the system output, the required power is supplied by the electric utility.

The inverter allows for the conversion of power and also shuts down the system during a power outage or other electrical disruption. This is a safety feature that prevents voltage from traveling back into the grid via backfeed.

3.3.0 Grid-Interactive Systems

Like a grid-tied system, a grid-interactive system is connected to the utility and uses inverted PV power as a supplement. Unlike a grid-tied system, which is literally tied to the grid and cannot function independently, a grid-interactive system provides a means of independent power. Grid-interactive systems include batteries that can supply power during outages and after sundown.

When the PV system operates, it first charges the batteries. It then satisfies the existing load, and any excess power is sent to the grid. As with off-grid systems, a charge controller is used to monitor the battery charge to ensure consistent power with minimal downtime. In the event of a power outage, battery power is supplied to critical loads through the use of a designated subpanel. Customers select the number of days of autonomy provided by the batteries. When these systems are used with other energy sources, such as wind turbines or generators, they are known as hybrid systems.

require power poles and transmission lines, and therefore eliminate the possibility of outages due to downed power lines. These are known as off-grid systems. Off-grid systems use batteries for energy storage as well as battery-based inverter systems. Charge controllers are used to maximize the battery-charging efficiency of the solar array. Backup power is provided by an engine-driven generator.

On Site

Emergency Shelters

Emergency shelters are starting to incorporate PV systems to ensure power in the event of a disaster. Depending on the location, government rebates and/or grants may be available to offset the costs of building qualified structures.

INVERTER AND
DC DISCONNECT

METER

PANEL

(AC DISCONNECT NOT SHOWN.)

74104-11_F15.EPS

Figure 15 Grid-tied PV system.

3.4.0 Utility-Scale Solar Generating Systems

A utility-scale PV system uses a large bank of solar cells to produce direct current (*Figure 16*). Motorized systems are often used to adjust the panel position so that it follows the movement of the sun throughout the day. This is known as tracking.

Utility-scale solar PV systems are not yet in common use and do not provide the same level of power as solar thermal power plants. However, it is expected that they will become increasingly popular as the technology advances.

74104-11_F16.EPS

Figure 16 Solar PV power plant.

4.0.0 POWER

Before you can understand how solar PV systems operate, you must first understand some basic concepts of power. Power is the ability to do work. Mechanical power is often expressed in horsepower (hp). In electrical circuits, power is measured in watts (W). One horsepower equals 746 watts. One watt is also the power used when one ampere (A) of current flows through a potential difference of one volt (V). For this reason, watts are referred to as volt-amps (VA). Voltage is also referred to as electromotive force (E). Power (P) is determined by multiplying the rated current (I) by the rated voltage (E):

$$P = I \times E$$

This equation is used to find the power consumed by a circuit or load when the values of current and voltage are known. Using variations of this equation, the power, voltage, or current in a circuit can be calculated whenever any two of the values are known. See *Figure 17*. Note that all of these formulas are based on the power formula (P = I × E).

Watt-hours (Wh) are calculated by multiplying the power in watts (VA) by the number of hours during which the power is used. The kilowatt-hour (kWh) is commonly used for larger amounts of electrical work or energy. (The prefix *kilo* means one thousand.) For example, if a light bulb uses 100W or 0.1 kW for 10 hours, the amount of energy consumed is 0.1 kW × 10 hours = 1.0 kWh.

Very large amounts of electrical work or energy are measured in megawatts (MW). (The prefix *mega* means one million.)

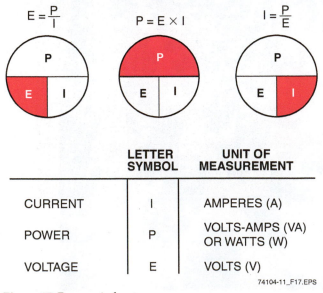

LETTER SYMBOL		UNIT OF MEASUREMENT
CURRENT	I	AMPERES (A)
POWER	P	VOLTS-AMPS (VA) OR WATTS (W)
VOLTAGE	E	VOLTS (V)

74104-11_F17.EPS

Figure 17 Power circle.

To determine the total panel or array output, the wattage is multiplied by the **peak sun hours** per day for the geographical area to determine the watt-hours produced per day. Peak sun hours or **insolation** values represent the equivalent number of hours per day when solar **irradiance** averages 1,000 W/m². Irradiance is a measure of radiation density and varies widely by location. Areas with few cloudy days and low levels of dust have high levels of irradiance. In addition, higher **elevations** have greater irradiance than those at **sea level**. Higher elevations also have cooler temperatures, which reduces the resistance and increases the array output voltage.

5.0.0 PV SYSTEM COMPONENTS

A typical grid-interactive PV system has four main components: panels, an inverter, batteries, and a charge controller. The remaining components are known as the **balance-of-system (BOS)** components. They include the panel mounts, wiring, overcurrent protection, grounding system, and disconnects. PV installations must comply with all applicable requirements of the *National Electrical Code*® (*NEC*®).

5.1.0 PV Panels

PV panels, sometimes referred to as modules, consist of numerous cells sealed in a protective laminate, such as glass. Solar cells work using a semiconductor that has been **doped** to produce two different regions separated by a p-n junction. Doping is the process by which impurities are introduced to produce a positive or negative charge. Crystalline silicon (c-Si) is used as the semiconductor in most solar cells.

PV panels are normally prewired and include positive and negative leads attached to a sealed termination box. Panels are rated according to their maximum DC power output in watts under **Standard Test Conditions (STC)**. Standard Test Conditions are as follows:

- Operating temperature of 77°F (25°C)
- Incident solar irradiance level of 1,000 W/m²
- **Air mass (AM)** of less than 1.5 **spectral distribution**

Spectral distribution is caused by the distortion of light through Earth's atmosphere. The air mass value is based on an ideal value of zero, which occurs in outer space.

Because panels are rated under ideal conditions, their actual performance is usually 85 to 90

Solar Energy

The amount of solar radiation that reaches Earth's outer atmosphere is nearly constant at 1,360 W/m². On a clear day, approximately 70 percent of this radiation travels through the atmosphere to Earth. Average values are typically 1,000 W/m² or lower, but can be as high as 1,500 W/m² when magnified by certain atmospheric conditions.

The peak sun hours for a given location represent an average value since solar intensity varies by time of day, season, and cloud cover. For example, a location may receive 800 W/m² for three hours and 1,200 W/m² for two hours. The total would be 4,800 W/m² divided by 1,000 W/m² = 4.8 peak sun hours.

To convert peak sun hours to power, multiply the wattage by the peak sun hours. For example, the power produced by a 36V/5A solar panel is 36V × 5A = 180W. If this 180W panel receives five peak sun hours per day, it produces 180W × 5 = 900 watt-hours (0.9 kWh) per day.

Note: Insolation maps and other resources can be found on the National Renewable Energy Laboratory (NREL) website at www.nrel.gov.

percent of the STC rating. The output of a PV cell depends on its efficiency, cleanliness, orientation, amount of sunlight, and temperature. Temperature increases can cause a significant decrease in the system output voltage. This is because higher air temperatures decrease irradiance. The irradiance drops by 50 percent for every 18°F temperature rise (10°C). (Note that this isn't a straight conversion because it represents an interval, rather than a defined temperature.)

PV cells are characterized by the type of crystal used in them. There are three basic types of PV cells: **monocrystalline**, **polycrystalline**, and **amorphous** (commonly known as **thin film**). The type of crystal used determines the efficiency of the cell. For example, a typical monocrystalline panel might produce 75W of power, while an equivalent polycrystalline cell might produce 65W, and a thin-film cell might produce 45W.

> **NOTE**
>
> Actual panel wattages vary widely depending on manufacturer, panel size, and type of semiconductor. Always consult the nameplate data for the specific panel in use.

5.1.1 Monocrystalline

Monocrystalline cells are formed using thin slices of a single crystal. Monocrystalline cells are currently the most efficient type of PV cell. Due to the manufacturing process, however, they are also the most expensive. A typical monocrystalline panel is shown in *Figure 18*.

5.1.2 Polycrystalline

Polycrystalline cells are made by pouring liquid silicon into blocks and then slicing it into wafers. This is a less expensive process than using a single crystal. However, poured silicon creates non-uniform crystals when it solidifies, reducing the efficiency of the panel. This gives these panels their characteristic flaked appearance (*Figure 19*).

74104-11_F18.EPS

Figure 18 Monocrystalline PV panel.

5.1.3 Thin Film

Thin-film PV panels are made using ultra-thin layers of semiconductor material. The reduced material use results in lower manufacturing costs, but also produces the lowest efficiency. Because of this, solar panels using thin-film cells must be larger to produce the same amount of energy as the other types.

Thin-film cells are often used in low-voltage applications, such as solar calculators and other electronics. They can also be encased in laminate to create rigid solar panels or used to create flexible building materials, such as roofing shingles (*Figure 20*). When a solar panel is built into a structure, it is known as **building-integrated photovoltaics (BIPV)**.

5.2.0 Inverters

Inverters are used to convert the DC produced by the PV array into AC that can be used by various loads (*Figure 21*). AC travels in a **sine wave**. There are two common types of inverters: modified sine wave and true sine wave. Modified sine wave in-verters are less expensive but do not provide the power quality of true sine wave inverters. They are not recommended for use with electronic equipment or other sensitive devices, including certain types of motors. True sine wave inverters produce smooth power similar to that provided by the grid. The quality of this power may actually be superior to grid power as it is free of voltage dips, spikes, and noise.

Inverters are available in a wide variety of types and load ratings. Some inverters are designed to operate with standalone systems, while others are designed for grid-tied and grid-interactive systems. Off-grid systems use battery-based inverter systems with an output voltage of 120/240 VAC so they can operate larger loads, such as water heaters and stoves. Inverters are rated in both continuous watts and surge watts. Some inverters include integral ground-fault protection and DC disconnects, while others require that these devices be installed separately. Many inverters use a digital display to indicate **ambient temperature**, voltage output, and other system parameters.

74104-11_F19.EPS

Figure 19 Polycrystalline PV panel.

THIN-FILM SOLAR PANEL

GT30E Grid Tie Inverter

STANDARD THREE-PHASE INVERTER

FLEXIBLE THIN-FILM USED IN
INTEGRATED ROOFING APPLICATION

74104-11_F20.EPS

Figure 20 Typical thin-film applications.

5.3.0 Batteries

Batteries are used to store energy produced by the PV array and supply it to electrical loads as needed (*Figure 22*). Batteries are also used to operate the PV array near its maximum power point, to stabilize output voltages during periods of low production, and to supply startup currents to loads such as motors.

Solar batteries require proper charging. While some deep cycle batteries can be discharged to 20 percent of capacity, it is best not to go that low as repeated deep cycling shortens battery life.

INVERTER WITH BATTERY SYSTEM

74104-11_F21.EPS

Figure 21 Inverters.

The amount of charge remaining is known as the **depth of discharge (DOD)**. Battery capacity is measured in amp-hours (Ah). For example, a 100 Ah battery from which 80 Ah has been withdrawn has undergone an 80 percent DOD. Some PV systems include a low-battery warning light or cutoff switch to prevent the damage caused by repeated deep cycling.

Figure 22 PV batteries.

Batteries must be protected from extreme temperatures. Cold temperatures reduce battery output while hot temperatures increase deterioration. With proper care and use, solar batteries can last between five and seven years. Batteries are rated by the expected number of charge cycles as well as the maximum discharge current. Each cell produces 2V, so a 6V battery contains three cells. Batteries can be connected in different ways to produce the desired output.

There are two types of solar batteries: flooded lead acid (FLA) and sealed absorbent glass mat (AGM).

5.3.1 FLA Batteries

FLA batteries are more common and less expensive than AGM batteries. However, they require special safety precautions to prevent accidental contact with the battery acid. They also require regular maintenance to add water and prevent corrosion. FLA batteries can deteriorate quickly if the fluid levels drop and the plates oxidize. They require good air circulation to prevent gases from accumulating to explosive levels or causing corrosion of other equipment. They are normally located in a separate compartment with appropriate venting and warning labels.

5.3.2 AGM Batteries

AGM batteries are sealed lead-acid batteries. The acid is contained in a special glass fiber mat so a vented battery compartment is not required. In addition, AGM batteries generally do not leak or produce corrosive gases. These batteries are also maintenance-free.

5.4.0 Charge Controllers

Charge controllers are used to regulate the charge and discharge of the system batteries. They are rated by their maximum AC current output, so it is essential that they be sized to match the application. There are two main types of charge controllers, including pulse width-modulated (PWM) and maximum power point tracking (MPPT). Older charge controllers use shunts or relay transistors and do not work well with sealed batteries. PWM charge controllers are ideal for use with sealed batteries but can only be used within a limited range of panel configurations and voltages. An MPPT charge controller is shown in *Figure 23*. These controllers harvest energy more efficiently and can be used with a wider range of array configurations and voltages. A digital display is provided to monitor power use, battery charge, and other system performance data. MPPT charge controllers are preferred for use in cold climates because they prevent the overvoltage that can occur at lower temperatures and provide more precise control over the DOD.

5.5.0 BOS Components

BOS components include the wiring, grounding system, disconnects, foundations, and support frames. They also include any required subpanels, conduit, and combiner boxes. Weatherproof combiner boxes are used to connect strings of so-

Figure 23 MPPT charge controller.

lar panels to create a larger array, and to provide a convenient array disconnect point.

5.5.1 Electrical System Components

A PV electrical system includes all wiring, devices, and components between the service entrance and the final panel termination. The most important safety devices in a PV system are the AC and DC disconnects and the grounding system.

Each piece of equipment in a PV system requires a switch or circuit breaker to disconnect it from all sources of power. A DC disconnect is shown in *Figure 24*. This disconnect is attached to the inverter. If the inverter does not include a DC disconnect, it must be purchased and installed separately.

The AC disconnect is located near the service panel. A typical AC disconnect is shown in *Figure 25*. PV disconnects may be located indoors if the panel is in the building.

The grounding system is essential to the electrical integrity of a PV array. All exposed non-current-carrying metal parts of the equipment must be bonded and connected to ground. A ground rod connection is shown in *Figure 26*.

Many ground-mounted PV systems use multiple levels of protection against stray voltages, such as lightning. These may include surge protectors and the use of grounded metal fencing for ground-mounted arrays.

Ground-fault protection is also required. Some inverters include ground-fault protection. If it is not supplied in the inverter, it will be located elsewhere in the PV system.

Other electrical system BOS components include the panel wiring, conduit, combiner boxes, and termination boxes. See *Figure 27*. All wiring must be sized for the load and the distance between components. All cables carrying medium or high voltages must be routed in conduit.

5.5.2 Footers and Support Structures

Support frames and footers are the most labor-intensive portion of any ground-mounted installation. In a system installed in an open field, the footers are typically concrete while the support frames are aluminum or steel.

Roof-mounted systems are mounted on aluminum or stainless steel rails secured into the roof trusses using lag bolts. See *Figure 28*. These supports include flashing to prevent water entry. Proper weathersealing of all roof penetrations is essential.

6.0.0 COLLECTING SOLAR DATA

Solar arrays work best when facing true solar south. True solar south is slightly different than a magnetic reference or compass south. A quick way to determine true solar south is to measure the length of time between sunrise and sunset, and then divide by two. The position of the sun at the resulting time represents true solar south.

Figure 24 DC disconnect.

74104-11_F24.EPS

Figure 25 AC disconnect.

74104-11_F25.EPS

Figure 26 Grounding system.

Figure 27 BOS components.

NCCER – *Alternative Energy* 74104-11

Figure 28 Roof-mounted rail support.

74104-11_F28.EPS

NOTE

Solar south can also be determined using the National Oceanic and Atmospheric Administration Sunrise/Sunset Calculator at www.srrb.noaa.gov.

PV systems are designed to maximize output by optimal placement in relation to the motion of the sun. The most important values include azimuth, altitude, and declination.

- *Azimuth* – For a fixed PV array, the azimuth angle is the angle clockwise from true north that the PV array faces. The default azimuth angle is 180° (south-facing) for locations in the northern hemisphere and 0° (north-facing) for locations in the southern hemisphere. This value maximizes energy production. In the northern hemisphere, increasing the azimuth favors afternoon energy production, while decreasing it favors morning energy production. The opposite is true for the southern hemisphere.

- *Altitude* – The altitude is the angle at which the sun is hitting the array. In many areas, panels are simply set to an angle that matches the local latitude. This provides a yearly average maximum output power. However, seasonal adjustments can increase the output in locations with high irradiance. In small systems, panels can be manually adjusted to optimize the tilt angle. Large systems are more likely to use automatic tilt adjustment.

- *Declination* – The declination of the sun is the angle between the equator and the rays of the sun. It ranges between +23.45° on the summer solstice (on or about June 21) to –23.45° on the winter solstice (on or about December 21). During the spring equinox (on or about March 21)

and the fall equinox (on or about September 21), the angle is zero. Because declination changes throughout the year, the optimal panel tilt angle also changes. This seasonal tilt of Earth causes longer shadows in winter because the sun is lower in the sky.

The Solar Pathfinder™ is a manual tool used to provide a full year of solar data for a specific location (*Figure 29*). It includes sun path diagrams for various latitude bands and an angle estimator for determining the sun's altitude and azimuth at various times of year. To use this instrument, you need to know the latitude and declination angle for the desired location. These can be found at the National Geophysical Data Center website at www.ngdc.noaa.gov. The latitude is used to select the correct band diagram and a compass is used to align the unit to the desired angle of declination. Shading is documented by tracing the trees in the reflection using a grease pencil. The PV system must be installed outside of the shaded area or trees must be removed to eliminate shading. Companion software is available that generates monthly sun paths for each specific site latitude instead of using the latitude band diagrams.

The Solmetric SunEye™ is an electronic device that allows users to instantly assess the total potential solar energy of a site (*Figure 30*). This device shows site shading in a digital display rather than through tracings. It is more expensive than the Solar Pathfinder™ but has the advantage in ease of use.

In most locations in the United States, winter produces the least amount of sunlight due to shorter days, increased cloud cover, and the sun's lower position in the sky. Insolation is usually highest in June and July and lowest in December and January. For best results, a solar PV site should be free of shade between 9 AM and 3 PM on December 21. *Figure 31* shows the path of the sun during the year for a specific location.

On Site

Solar Calculators

Online solar calculators are available to simplify sizing calculations. For example, go to www.findsolar.com and plug in the zip code, monthly power usage, and desired percentage of solar power. It will provide general sizing information for the selected PV system.

Measuring Irradiation

Pyranometers are used to measure irradiance across the entire sky. Some units are mounted at the same angle as the array. They provide long-term data that can be compared against the array output. Other pyranometers are handheld and plug into a digital reader. Handheld units are less precise but offer the advantages of portability and rapid field measurements.

MOUNTED PYRANOMETER

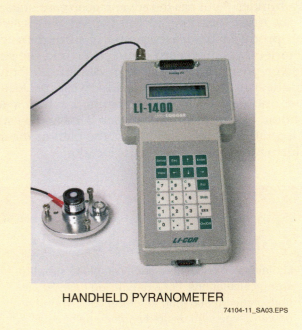

HANDHELD PYRANOMETER

74104-11_SA03.EPS

NOTE

You may need to adjust for Daylight Savings Time. See the instructions provided for the instrument in use.

For example, to find the approximate altitude and azimuth of the sun in April, follow along on the curve that reads Apr. & Aug. 21 until you find 2:30 PM. The altitude is 45° while the azimuth is 60°. Note that values outside of the lines must be estimated.

7.0.0 INSTALLATION

The two most common system installations are roof-mounted and ground-mounted arrays. Roof-mounted installations are more common in residential work where the array space on the ground may be limited due to lot size. In addition, roof-mounted arrays help to limit accidental contact with energized components. Ground-mounted installations are often used for commercial systems where access can be limited by restricted areas or fencing. A ground-mounted installation is more time-intensive than a roof installation due to the larger support system.

7.1.0 Roof-Mounted Installations

Panels are secured to the roof deck in one of three ways: direct mount, rack mount, and standoff mount:

- *Direct mount* – Direct mount systems mount directly to the roof. These systems provide an unobtrusive appearance and are subjected to the lowest wind loads. However, they prohibit air circulation beneath the panel. This increases the panel temperature, decreasing the output. *Figure 32* shows a direct-mount roof array.
- *Rack mounts* – Rack mounts secure the panels using a triangular support. They may be supplied with or without an adjustable tilt angle. They are commonly used on flat roofs and designed to match the desired tilt angle. These mounts provide the greatest degree of air circulation and also provide easy access to electrical connections. However, they are also subject to the greatest degree of wind loads. When installing rack-mounted arrays, it is essential to follow the manufacturer's instructions regarding inter-row spacing. Improperly spaced arrays create shadows, seriously reducing or even preventing system output voltages.

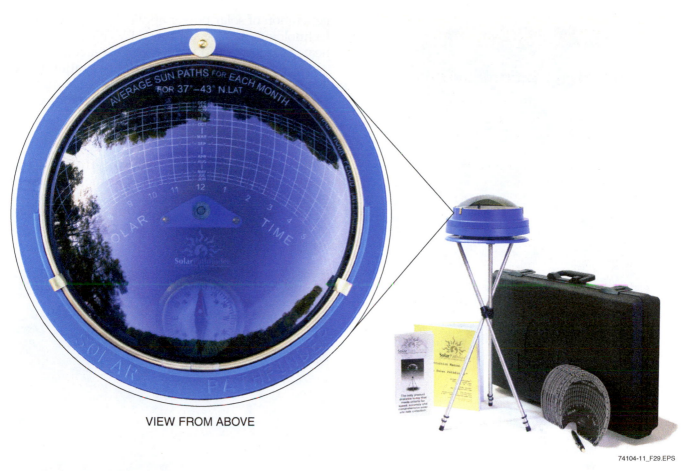

VIEW FROM ABOVE

74104-11_F29.EPS

Figure 29 Solar Pathfinder™.

- *Standoff mounts* – Standoff mounts are the most common type of mounting system for sloped roofs. They provide three to five inches of space between the panel and the roof. This allows for good air circulation while minimizing wind loads.

> **NOTE**
>
> Some commercial rooftop systems use a ballast (weighted) system rather than traditional roof mounts. Due to weight and wind loading, these systems must be engineered to match the specific structure and location.

7.2.0 Ground-Mounted Installations

When installing a ground-mounted array, the site must first be cleared of any vegetation and leveled. Next, concrete footers are installed to support the array. They may be precast (*Figure 33*) or poured in place. This anchor system requires engineering assistance to ensure sufficient protection against expected loads.

After the footers are installed, the support system is erected according to the manufacturer's in-structions (*Figure 34*). *Figure 35* shows a completed support system.

When the support system is complete, the panels are installed. All fasteners must be tightened to the manufacturer's specifications (*Figure 36*). *Figure 37* shows a completed ground-mounted array. White marble chips or other light-colored stone is used to reflect light and maximize panel output. *Figure 38* shows the electrical portion of the installation.

8.0.0 US DOE Position

According to the Department of Energy (DOE), its Solar Energies Technologies Program "accelerates the development of solar technologies as energy sources for the nation and world. The solar program also educates the public about the value of solar as a secure, reliable, and clean energy choice." Both the demand for more energy and current environmental challenges have emphasized the benefits of clean, abundant solar energy.

One of the DOE's missions is to encourage the nation's solar energy research and development program. The focus is on strengthening basic technology and aiding the introduction of the next generation of solar power applications. Developing technologies include advanced PV cells, solar water heaters, and concentrating solar power collectors.

The Energy Information Administration (EIA) tracks statistical information on the production and use of solar thermal and solar PV energy.

9.0.0 EMERGING TECHNOLOGIES

There are tremendous research efforts in the field of PV systems. Some of the emerging areas of research include replacements for traditional semiconductors, while others involve methods of improving efficiency in solar collection and energy storage.

74104-11_F30.EPS

Figure 30 Solmetric SunEye™.

74104-11_F32.EPS

Figure 32 Installed roof array.

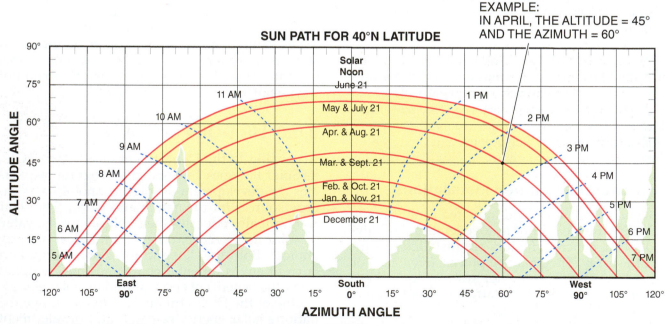

EXAMPLE:
IN APRIL, THE ALTITUDE = 45°
AND THE AZIMUTH = 60°

To use this chart for southern latitudes, reverse the horizontal axis (east/west & AM/PM).

74104-11_F31.EPS

Figure 31 Sun path diagram.

Figure 33 Setting footers.

Figure 35 Completed support system installation.

Figure 34 Installing the support system.

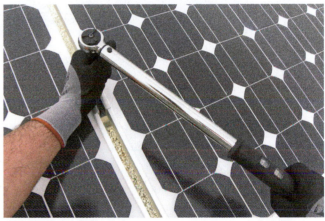

Figure 36 Torquing fasteners.

One new type of system replaces silicon with a light-sensitive dye that absorbs light and produces current. These are known as **electrochemical solar cells**. When this technology matures, electrochemical cells may be very popular as they are simple and less expensive than traditional solar cells.

Research in solar energy storage includes the development of household **fuel cells** to replace traditional battery systems. These differ from traditional fuel cells, which use rare components and are more expensive to manufacture. This process uses sunlight and a special chemical component known as a catalyst to split water into hydrogen and oxygen, which are later recombined in the fuel cell to produce energy. It is efficient and emission free. According to researchers at the Massachusetts Institute of Technology (MIT), it is possible that household fuel cells could begin to replace wire-based energy delivery within the next decade.

Figure 37 Installed ground-mounted array.

Microinverters

Enphase microinverters attach directly to individual solar panels using low-voltage DC wiring. This increases system safety and allows for easy expansion.

74104-11_F38.EPS

Figure 38 Electrical components.

SUMMARY

Solar power provides a clean, renewable alternative to fossil fuels. It can be harnessed without disrupting the environment and produces no hazardous waste or emissions. It is also quiet and requires little maintenance. In addition, utility and government rebates are available to offset the cost of installing these systems.

Solar thermal systems use sunlight to heat fluids or spaces. When sunlight is used directly to produce heat, it is known as passive solar power. Solar thermal systems can be used in low-temperature, medium-temperature, and high-temperature applications. Low-temperature systems are used for applications such as space heating or as solar pool heaters. Medium-temperature systems are often used to heat water for home use. High-temperature solar thermal systems are used in utility-scale solar thermal generating systems.

Solar photovoltaic (PV) systems convert sunlight into electricity. When sunlight is used to produce another form of energy, such as electricity, it is known as active solar power. Solar PV cells produce electricity through the use of semiconductors. Many cells are contained in a solar panel and these panels are connected in an array.

Solar PV can be classified into four types of systems: standalone systems, grid-tied systems, grid-interactive systems, and utility-scale solar generating systems. Most residential and commercial customers use grid-tied and grid-interactive systems. These systems provide an advantage in that power is always available and excess power can be sold back to the utility in the form of credits. In standalone and grid-interactive systems, backup power is provided by a system of one or more batteries. These batteries are matched to the appropriate charge controller to provide grid-quality power.

As this technology continues to develop, the use of solar power systems will become more prevalent.

Review Questions

1. Which of the following is an active form of solar power?

   a. Designing a building to maximize sunlight exposure in winter
   b. Hanging clothes on a line
   c. Heating water in an outdoor tank
   d. Using a solar-powered calculator

2. When excess energy is sent back into the grid, it is known as _____.

   a. buyback metering
   b. grid metering
   c. net metering
   d. carryover metering

3. The use of solar power is now increasing by _____.

   a. 5 percent per year
   b. 10 percent per year
   c. 15 percent per year
   d. 25 percent per year

4. An example of a low-temperature solar thermal system would be a _____.

   a. pool heater
   b. solar boiler
   c. power tower
   d. evacuated tube system

5. Which of the following is likely to use a dual-axis tracking system?

   a. A solar concentrator system
   b. A residential off-grid system
   c. A small commercial grid-tied system
   d. A residential grid-interactive system

6. A solar boiler is a type of _____.

   a. grid-tied system
   b. grid-interactive system
   c. utility-scale system
   d. standalone system

7. When a motorized system is used to adjust a PV array to follow the motion of the sun, it is known as _____.

   a. trailing
   b. trolling
   c. tracing
   d. tracking

8. One horsepower is equal to _____.

   a. 1 watt
   b. 76 watts
   c. 144 watts
   d. 746 watts

9. If an electric heater uses 1,200W for 10 hours, the power consumed is _____.

   a. 1.2 kWh
   b. 12 kWh
   c. 1,200 kWh
   d. 12,000 kWh

10. The prefix *kilo* means _____.

   a. 10
   b. 100
   c. 1,000
   d. 1,000,000

11. Which of the following locations is likely to have the highest irradiance?

   a. A desert area with significant dust but few cloudy days
   b. A mountaintop with cool temperatures but no dust and few cloudy days
   c. A hot southern location at sea level
   d. A cool but cloudy northern location

12. A type of PV panel having a characteristic flaked appearance is probably made up of _____.

   a. monocrystalline cells
   b. megacrystalline cells
   c. polycrystalline cells
   d. amorphous cells

13. The type of PV cell used in BIPV applications is _____.

   a. monocrystalline cells
   b. megacrystalline cells
   c. polycrystalline cells
   d. thin film

14. An advantage in choosing FLA batteries over AGM batteries is that FLA batteries _____.

   a. are less expensive
   b. require no maintenance
   c. do not leak
   d. do not have special venting requirements

15. The best charge controller for use in a cold climate is a _____.

    a. MPPT charge controller
    b. PWM charge controller
    c. shunt charge controller
    d. relay transistor charge controller

16. Suppose the sun rises at 6:00 AM and sets at 9:00 PM. What time should you note the position of the sun in order to determine true solar south?

    a. 6:00 AM
    b. 12:00 PM
    c. 1:30 PM
    d. 3:00 PM

17. The default azimuth angle for locations in the southern hemisphere is _____.

    a. 0 degrees
    b. 90 degrees
    c. 180 degrees
    d. 270 degrees

18. During the spring equinox, the angle of declination is _____.

    a. zero
    b. +23.45 degrees
    c. –23.45 degrees
    d. 90 degrees

19. A flat roof is most likely to use a _____.

    a. direct mount system
    b. building-integrated system
    c. rack mount system
    d. standoff mount system

20. Which government body tracks statistical data related to solar PV/thermal systems?

    a. Bureau of Economic Analysis (BEA)
    b. Energy Information Administration (EIA)
    c. Department of Energy (DOE)
    d. Department of Agriculture (USDA)

Name: _____

Date: _____

74104-11_CW01A.EPS

Across:

2. A group of sun-tracking mirrors used to focus the sunlight on a central receiver in a solar thermal power plant.

4. A low-efficiency type of photovoltaic cell that can be used in flexible forms; also known as thin-film.

7. A temporary decrease in grid output voltage typically caused by peak load demands.

8. The generic term for solar insolation.

9. An array mounting system designed to adjust either the horizontal or the vertical axis of a panel to follow the movement of the sun.

11. A power production system that converts sunlight into electricity using a semiconductor.

12. A material that exhibits the properties of both a conductor and an insulator.

15. The equivalent number of hours per day when solar irradiance averages 1,000 W/m^2.

18. A unit of energy typically used for metering power.

20. A synonym for amorphous.

21. A complete PV power generating system including solar panels.

22. A tracking system designed to adjust both the horizontal and vertical axes of a panel to precisely follow the movement of the sun.

24. A device that harnesses the energy from a chemical reaction between hydrogen and oxygen to produce direct current.

27. A material to which specific impurities have been added to produce a positive or negative charge.

28. A method of determining a location on the Earth in reference to the equator.

32. The distortion of light through Earth's atmosphere results from this.

34. Another term for a grid-connected system.

35. A grid-interactive system used with other energy sources, such as wind turbines or generators.

38. A type of solar thermal system consisting of a tank mounted above one or more flat panel collectors that relies on convection to circulate water.

39. Describes large solar farms designed to produce power in quantities large enough to operate a small city.

42. The atmosphere that solar radiation must pass through to reach Earth.

44. A type of battery charge controller that provides precise charge/discharge control over a wide range of battery temperatures.

46. A junction box used to connect strings of solar panels to create a larger array, and to provide a convenient array disconnect point.

48. A highly efficient PV cell formed using thin slices of a single crystal.

50. The panel support system, wiring, disconnects, and grounding system that are installed to support a PV array.

51. A method of measuring power used from the grid against PV power put into the grid.

52. A PV system that operates in parallel with the utility grid and provides supplemental power to the building or residence.

Down:

1. The angle between the equator and the rays of the sun.

3. A type of collector that maximizes the gathering of solar energy by using mirrors or lenses to focus sunlight onto a central receiver.

5. A device used to regulate the charging of the battery system to prevent overcharge and excess discharge.

6. A type of solar thermal collection system consisting of one or more storage tanks in an insulated box with a glazed side facing the sun.

10. A PV system that supplies supplemental power and can function independently through the use of a battery bank.

12. The sun's altitude and azimuth at various times of year for a specific location or latitude band.

13. A type of PV cell formed by pouring liquid silicon into blocks and then slicing it into wafers.

14. A type of solar heat collector consisting of parallel rows of transparent glass tubes with a vacuum between the tubes to preserve heat.

16. The angle at which the sun is hitting the array.

17. A set of consistent circumstances, such as temperature, solar irradiance, etc., used to test solar panels for comparison.

19. A measure of the average height of the ocean's surface between low and high tide.

21. The air temperature of an environment.

23. A PV system used to provide power in remote areas, typically using batteries and possibly other sources of power other than the grid.

25. A measure of a location's relative height in reference to sea level.

26. A system that uses sunlight to heat air or water.

29. A synonym for an off-grid system.

30. A type of solar heat collector consisting of a metal box with a transparent cover containing three main components: an absorber to collect heat, tubes with water or air, and insulation.

31. A PV system built into the structure replacing a building component, such as roofing.

33. A type of PV cell that replaces silicon with a light-sensitive dye that absorbs light and produces current.

36. A device used to convert direct current to alternating current.

37. A measure of the amount of charge removed from a battery system.

38. The position of a panel or array in reference to a horizontal facing.

40. The number of days a fully charged battery system can supply power to loads without recharging.

41. A modulating control that uses a rapid switching method to simulate a waveform and provide smooth power.

43. A form presented as a ripple, representing the consistent frequency and amplitude of electrical power.

45. A measure of radiation density at a specific location.

47. Current flowing into the grid instead of out.

48. A PV system component consisting of numerous connected PV cells encased in a protective glass or laminate frame. Also known as a PV panel.

49. For a fixed array, the angle clockwise from true north that the PV array faces.

Michael J. Powers

Corporate Safety Training Director
Tri-City Electrical Contractors, Inc.
Altamonte Springs, FL

Mike Powers has done everything in energy from wiring a kennel in the midst of a pack of barking dogs to working on solar panels. He has had a varied and rewarding career that has kept him interested and enjoying life.

Who inspired you to enter the industry?
You could say the smell of hot grease inspired me. My first job was flipping hamburgers in a fast food restaurant, and it didn't take long till I was ready for a change. My dad was an electrician, and that looked like a much better career choice.

Tell us about your apprenticeship experience.
The program I apprenticed in was a good one. The electricians I worked under had a broad range of experience, so I was prepared for many different situations. I worked all over, in everything from kennels to colleges. I even worked in a fast-photo print shop once.

What positions have you held and how did they help you to get where you are now?
After I completed my apprenticeship, I became a licensed electrician, a job-site superintendent, a master electrician, and I'm currently a corporate safety and training director. The electrical theory taught in apprenticeship school made a big difference. I learned a lot studying for my licensing exams—those things gave me a good understanding of not just how but why certain things were done. Then there was the practical experience of being on the job for over 30 years and seeing the kinds of things that were needed. These different kinds of experiences gave me a good foundation for my current job in training. I'm now on the solar photovoltaic panel for NCCER's new program. It's stimulating to be in on something new and be part of a leading-edge technology.

What do you enjoy most about your work?
Problem-solving and people. Both are interesting and both can be a challenge.

Do you think training and education are important in construction?
If you want to have a career, it's not a choice. Our industry is changing and becoming more demanding every day, and you won't get anywhere standing still. The more you learn, the better you'll be prepared when an opportunity comes along.

Have you ever witnessed an accident that better training might have prevented?
I watched a carpenter step into a hole in a deck on the second floor. He stepped onto a 4×4 which he assumed was supported from below by a jack. He was not wearing any fall protection and the wood was toenailed in place, not braced. The board fell and he went with it. He suffered serious injuries. Because of the sand on the job site, several dozen workers had to push the ambulance to the building to take him to the hospital. Better training—and stricter enforcement of safety rules—might have made a difference.

How has training impacted your life and your career?
It's made me confident that I can do whatever comes my way. And it's given me the chance to move into new positions, make a better living, and expand my interests. It's kept me from getting stale.

How do you define craftsmanship?
Just like I define character—it's doing your job correctly, to the best of your ability, in a safe, productive manner, whether anyone else knows about it or not. It's personal.

Trade Terms Introduced in This Module

Air mass: The thickness of the atmosphere that solar radiation must pass through to reach Earth.

Altitude: The angle at which the sun is hitting the array.

Ambient temperature: The air temperature of an environment.

Amorphous: A low-efficiency type of photovoltaic cell characterized by its ability to be used in flexible forms. Also known as thin film.

Array: A complete PV power generating system including panels, inverter, batteries and charge controller (if used), support system, and wiring.

Autonomy: The number of days a fully charged battery system can supply power to loads without recharging.

Azimuth: For a fixed PV array, the azimuth angle is the angle clockwise from true north that the PV array faces.

Backfeed: When current flows into the grid.

Balance of system (BOS): The panel support system, wiring, disconnects, and grounding system that are installed to support a PV array.

Brownout: A temporary decrease in grid output voltage typically caused by peak load demands.

Building-integrated photovoltaics (BIPV): A PV system built into the structure as a replacement for a building component such as roofing.

Charge controller: A device used to regulate the charging and discharging of the battery system to prevent overcharge and excess discharge.

Combiner box: A junction box used to connect strings of solar panels to create a larger array, and to provide a convenient array disconnect point.

Concentrating collector: A device that maximizes the collection of solar energy by using mirrors or lenses to focus sunlight onto a central receiver.

Declination: The angle between the equator and the rays of the sun.

Depth of discharge (DOD): A measure of the amount of charge removed from a battery system.

Doped: A material to which specific impurities have been added to produce a positive or negative charge.

Dual-axis tracking: An array mounting system designed to adjust both the horizontal and vertical axes of a panel to precisely follow the movement of the sun.

Electrochemical solar cell: A type of PV cell that replaces silicon with a light-sensitive dye that absorbs light and produces current.

Elevation: A measure of a location's relative height in reference to sea level.

Evacuated tube collector: A type of solar thermal heat collector consisting of parallel rows of transparent glass tubes with a vacuum between the tubes to preserve heat.

Flat plate collector: A type of solar thermal heat collector consisting of metal box with a transparent cover that contains three main components: an absorber to collect the heat, tubes containing water or air, and insulation.

Fuel cell: A device that harnesses the energy produced by a chemical reaction between hydrogen and oxygen to produce direct current.

Grid-connected system: A PV system that operates in parallel with the utility grid and provides supplemental power to the building or residence. Since they are tied to the utility, they only operate when grid power is available. Also known as a grid-tied system.

Grid-interactive system: A PV system that supplies supplemental power and can also function independently through the use of a battery bank that can supply power during outages and after sundown.

Grid-tied system: See *grid-connected system*.

Heliostats: A group of sun-tracking mirrors used to focus the sunlight on a central receiver in a solar thermal power plant.

Hybrid system: A grid-interactive system used with other energy sources, such as wind turbines or generators.

Insolation: The equivalent number of hours per day when solar irradiance averages 1,000 W/m^2. Also known as peak sun hours.

Integral-collector storage system: A type of solar thermal system consisting of one or more storage tanks in an insulated box with a glazed side facing the sun.

Inverter: A device used to convert direct current to alternating current.

Irradiance: A measure of radiation density at a specific location.

Latitude: A method of determining a location on Earth in reference to the equator.

Maximum power point tracking (MPPT): A battery charge controller that provides precise charge/discharge control over a wide range of temperatures.

Module: A PV system component consisting of numerous electrically and mechanically connected PV cells encased in a protective glass or laminate frame. Also known as a PV panel.

Monocrystalline: A type of PV cell formed using thin slices of a single crystal and characterized by its high efficiency.

Net metering: A method of measuring power used from the grid against PV power put into the grid.

Off-grid system: A PV system typically used to provide power in remote areas. Off-grid systems use batteries for energy storage as well as battery-based inverter systems. Also known as a standalone system.

Peak sun hours: See *insolation*.

Polycrystalline: A type of PV cell formed by pouring liquid silicon into blocks and then slicing it into wafers. This creates non-uniform crystals with a flaked appearance that have a lower efficiency than monocrystalline cells.

Pulse width-modulated (PWM): A control that uses a rapid switching method to simulate a waveform and provide smooth power.

Sea level: A measure of the average height of the ocean's surface between low and high tide. Sea level is used as a reference for all other elevations on Earth.

Semiconductor: A material that exhibits the properties of both a conductor and an insulator.

Sine wave: A form presented as a ripple, representing the consistent frequency and amplitude of electrical power.

Single-axis tracking: An array mounting system designed to adjust either the horizontal or the vertical axis of a panel to follow the movement of the sun.

Solar photovoltaic (PV) system: A power production system that converts sunlight into electricity using a semiconductor.

Solar thermal system: A system that uses sunlight to heat air or water.

Spectral distribution: The distortion of light through Earth's atmosphere.

Standalone system: See *off-grid system*.

Standard Test Conditions (STC): Standardized panel ratings based on a specific operating temperature, solar irradiance, and air mass.

Sun path: The sun's altitude and azimuth at various times of year for a specific location or latitude band.

Thermosiphon system: A type of solar thermal system consisting of a tank mounted above one or more flat panel collectors. Thermosiphon systems rely on convection to circulate water through the collectors and to the tank.

Thin film: See *amorphous*.

Tilt angle: The position of a panel or array in reference to horizontal. Often set to match local latitude or in higher-efficiency systems, the tilt angle may be adjusted by season or throughout the day.

Utility-scale solar generating system: Large solar farms designed to produce power in quantities large enough to operate a small city.

Watt-hours (Wh): A unit of energy typically used for metering.

Additional Resources

This module presents thorough resources for task training. The following resource material is suggested for further study.

National Electrical Code® (NFPA 70), Latest Edition. National Fire Protection Association (NFPA): Quincy, MA.

Photovoltaic Systems, Second Edition. James P. Dunlop. Orland Park, IL: American Technical Publishers.

Solar Water Heating, Second Edition. Benjamin Nusz. Gabriola Island, BC, Canada: New Society Publishers.

Uniform Solar Energy Code, Latest Edition. Ontario, CA: International Association of Plumbing and Mechanical Officials (IAPMO).

Figure Credits

Courtesy of DOE/NREL, Module opener, SA02, and Projects 1-3

NASA, SA01

Mike Powers, Figures 2, 5, 24–27, and 33–38

© 2011 Photos.com, a division of Getty Images. All rights reserved., Figures 3 and 6

Sharp USA, Figures 4, 15, 18, 19 (right photo), 20 (top photo), and 32

©iStockphoto.com/Dovapi, Figure 7

©iStockphoto.com/alexandr6868, Figure 9

©iStockphoto.com/sharifphoto, Figure 10

Stirling Energy Systems, Figure 11

eSolar Inc., Figure 12

Antonio Vazquez, Figures 13, 14, and 28

Nellis Air Force Base, Figure 16

Topaz Publications, Inc., Figures 19 (left photo) and 22

Photo courtesy of Energy Conversion Devices, Inc. & United Solar Ovonic LLC, Figure 20 (bottom photo)

Schneider Electric, Figure 21

Outback Power Systems, Figure 23

The Eppley Laboratory, SA03 (top photo)

Copyright LI-COR, Inc. and used by permission, SA03 (bottom photo)

Solar Pathfinder, Figure 29

With permission of Solmetric Corporation, Figure 30

NCCER CURRICULA — USER UPDATE

NCCER makes every effort to keep its textbooks up-to-date and free of technical errors. We appreciate your help in this process. If you find an error, a typographical mistake, or an inaccuracy in NCCER's curricula, please fill out this form (or a photocopy), or complete the online form at **www.nccer.org/olf**. Be sure to include the exact module ID number, page number, a detailed description, and your recommended correction. Your input will be brought to the attention of the Authoring Team. Thank you for your assistance.

Instructors – If you have an idea for improving this textbook, or have found that additional materials were necessary to teach this module effectively, please let us know so that we may present your suggestions to the Authoring Team.

NCCER Product Development and Revision

13614 Progress Blvd., Alachua, FL 32615

Email: curriculum@nccer.org
Online: www.nccer.org/olf

❑ Trainee Guide ❑ AIG ❑ Exam ❑ PowerPoints Other _____

Craft / Level: _____ Copyright Date: _____

Module ID Number / Title: _____

Section Number(s): _____

Description: _____

_____

_____

_____

Recommended Correction: _____

_____

_____

_____

_____

Your Name: _____

Address: _____

_____

Email: _____ Phone: _____

74105-11

Wind Power

Module Five

Trainees with successful module completions may be eligible for credentialing through NCCER's National Registry. To learn more, go to **www.nccer.org** or contact us at **1.888.622.3720**. Our website has information on the latest product releases and training, as well as online versions of our *Cornerstone* newsletter and Pearson's product catalog.

Your feedback is welcome. You may email your comments to **curriculum@nccer.org,** send general comments and inquiries to **info@nccer.org**, or use the User Update form at the back of this module.

Copyright © 2011 by the National Center for Construction Education and Research (NCCER) and published by Pearson Education, Inc., publishing as Prentice Hall. All rights reserved. Manufactured in the United States of America. This publication is protected by Copyright, and permission should be obtained from NCCER prior to any prohibited reproduction, storage in a retrieval system, or transmission in any form or by any means, electronic, mechanical, photocopying, recording, or likewise. To obtain permission(s) to use material from this work, please submit a written request to NCCER Product Development, 13614 Progress Blvd., Alachua, FL 32615.

74105-11
WIND POWER

Objectives

When you have completed this module, you will be able to do the following:

1. List the advantages and disadvantages of wind energy.
2. Describe the past, present, and future of wind energy.
3. Describe wind power, how it is harnessed, and how it is used to generate energy.
4. Identify and describe wind energy applications.

Performance Tasks

This is a knowledge-based module; there are no performance tasks.

Trade Terms

Anemometer
Betz limit
Charge controller
Dynamo
Effective ground level
Furling
Horizontal-axis wind turbines (HAWTs)
Inverter
Kinetic energy
Longitudinal
Nacelle

Net metering
Pitch control
Power density
Supervisory control and data acquisition (SCADA) system
Swept area
Tip brakes
Vertical-axis wind turbines (VAWTs)
Wind rose
Wind shear
Yaw control

Industry Recognized Credentials

If you're training through an NCCER-accredited sponsor you may be eligible for credentials from NCCER's Registry. The ID number for this module is 74105-11. Note that this module may have been used in other NCCER curricula and may apply to other level completions. Contact NCCER's Registry at 888.622.3720 or go to nccer.org for more information.

Contents

Topics to be presented in this module include:

Figures and Tables ──────────

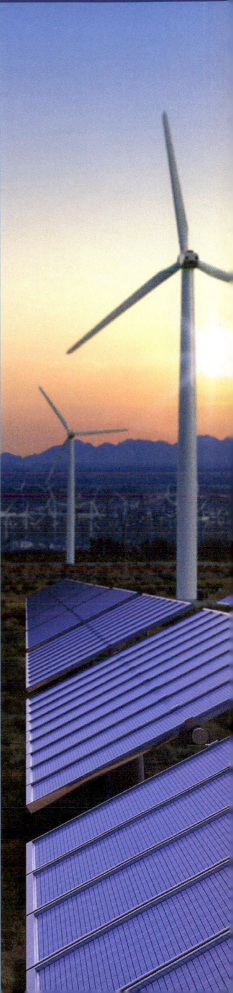

1.0.0 INTRODUCTION

There are many ways and reasons to put wind to work for us. Although research and new discoveries will likely improve the ability to harness wind energy, the generation of power from the wind (*Figure 1*) is practical and fully renewable technology. Unless you live in an area where the existence of wind turbines is obvious, you may not know how far it has already progressed. Billions are invested in wind power every year. The United States is presently on course to generate up to 20 percent of its electrical power from wind by 2030. Here are some statistics that may surprise you:

- By the end of 2009, US wind power capacity had grown to more than 35,000 megawatts (MW). This is roughly equal to 100 common coal-fired power plants. In 2009 alone, nearly 10,000 MW of capacity was added. A megawatt is equal to 1,000,000 watts. Ten car engines create about this much power. It is estimated that one megawatt is enough to power 300 average homes.
- From 2004 to 2008, the United States led the world in annual additions to wind power capacity. China led this category in 2009.
- Denmark, Spain, Portugal, and Ireland provide more than 10 percent of their total electrical power from wind turbines. Denmark led the way at 20 percent.
- At the end of 2009, wind capacity provided just over 2.5 percent of US national annual needs. Iowa presently generates nearly 20 percent of its power from wind.

There is little doubt that wind is one of the most popular resources. Not only can people benefit from the power, the wind industry provides economic growth opportunities. Wind energy is a resource that can reduce the need for fossil fuels.

The advantages and disadvantages of wind power are well known now. Here are just a few of the advantages to be considered:

- There is a lot of capacity available on a worldwide scale; wind is an inexhaustible resource.
- Wind is a clean source of energy. There are no atmospheric emissions beyond the process of making the needed parts.
- Wind energy can help reduce dependence on fossil fuels and those who control them.
- The wind is free; with modern technology, it can be captured efficiently.
- Although utility-scale wind turbines can be very tall, each takes up only a small plot of land. This allows the land below to be used for agriculture or other appropriate purposes.
- Many people find that wind farms add an interesting and unique feature to the landscape.
- Remote or undeveloped areas that are not connected to the power grid can use wind turbines to produce local power.

To be fair, some disadvantages should be acknowledged. Even supporters of the technology may qualify their support by adding, "as long as it's not in my backyard!" These words have become a part of modern culture, often referred to by the acronym NIMBY.

- Wind farms are not visually appealing to many people. Land-based wind farms cover large areas and are difficult to hide due to their height. Many people prefer the environment to retain its natural landscape.
- The persistent low-frequency whirring from the rotor can be annoying. Some minor noise can sometimes be heard from other components as well. Worn brake pads and load-bearing surfaces can make noise as the **nacelle** pivots atop the tower.
- Shadow flickers and light reflection from the blades as they rotate can be a visual nuisance.
- Ice formations on blades can be dangerous, traveling a significant distance if they suddenly break loose.
- Birds and bats may be killed by the blades.
- The wind is inconsistent in direction or intensity, varying from calm to storm force. As a result, there are times when turbines produce no power at all. This can lead to instability in the power grid.

Every aspect of wind power and the use of turbines will continue to be debated for years to come. However, a commitment to its use on a global scale is very real and well under way.

Figure 1 Wind turbine.

74105-11_F01.EPS

This module introduces wind power as a whole, from its history to the components that make it possible.

2.0.0 WIND POWER – PAST, PRESENT, AND FUTURE

There is certainly nothing new about the wind—it can be assumed that it predates human presence. Harnessing the power in wind does have a relatively well-documented history. Just as early man discovered valuable uses for the sun, the wind proved to be a resource as well. Of course, many years passed before man discovered the necessary technology to develop electrical power using the wind's energy. Today, technology seems to advance at the speed of light. The evolution of wind technology can be seen in *Figure 2*. Innovation and simplification are evident in new designs, while the size of commercial- and utility-scale turbines continues to grow. The future of wind power can only become brighter.

2.1.0 Origin and History of Wind Power

A lot of evidence exists to prove that wind power was used to propel boats as early as 5,000 BC. More than 2,000 years ago, the Chinese used wind power to drive water pumps. During the same period in other areas of the world, wind power was used to grind grain. Although communication was primarily limited to word of mouth from traveling merchants, reports of the growing uses for wind power were shared everywhere.

History suggests that in Western Europe, the horizontal-axis structure of the water wheel was the inspiration for early windmills. Post mills in Europe seemed more advanced than earlier wind-powered systems in Persia. Thus they were not obvious descendants of that approach. Hand-made wooden gears were made to transmit power from the horizontal shaft to a vertical shaft. The mill structure itself was balanced on a center post. This allowed the entire assembly to be rotated into the wind by the miller or by animal power.

Tower mills (*Figure 3*) used the post-mill concept in a more practical way. The lower portion of a tower mill remained fixed in position, often providing housing for the operators. The upper portion functioned as a post mill, with the ability to rotate into the wind as required. These structures were built primarily of brick or stone instead of wood.

Over the years, uses for wind power were found in many applications. Uses included saw-

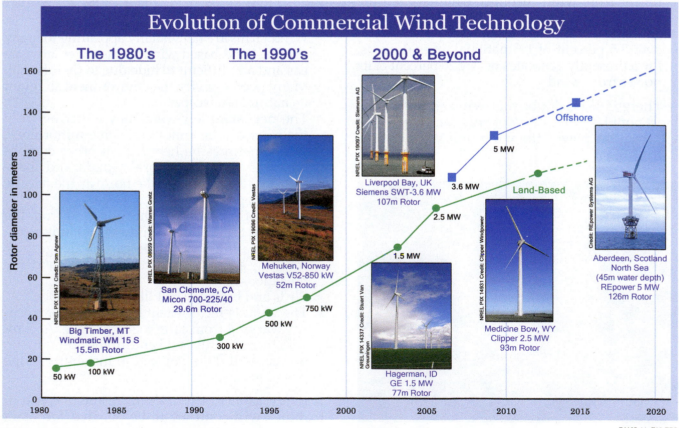

Figure 2 The evolution of commercial wind technology.

Figure 3 Tower mill.

ing, spice and cocoa processing, and the mixing of paints and dyes. If a process could benefit from a slow and powerful shaft turning, wind power could be used.

Industrialization and the development of steam power slowed developments in wind power for quite some time. There were obvious advantages to power production on demand, rather than being at the mercy of the wind. Technology, however, did lead to advanced blade designs and more efficient use of the wind to produce work.

A rapidly advancing industrial society stalled the advancement of wind power somewhat. This progress also created a hunger for more electrical power. The first windmill built specifically for power generation was erected in 1888 in Cleveland, Ohio by Charles Brush (*Figure 4*). A wealthy man due to more than 50 patented inventions, Brush founded the Brush Electric Company. The company provided power for a number of public and private entities. His company eventually became a part of the General Electric Company.

This first electrical turbine in the United States (*Figure 4*) was built on the site of the Brush mansion. It had 144 blades with a diameter over 17 meters (56 feet). The tail was more than 18 meters (60 feet) long and roughly 6 meters (20 feet) wide. At its best, the turbine was able to spin a **dynamo** at

Figure 4 Charles Brush and his wind turbine.

about 500 revolutions per minute (rpm). It used a 50:1 step-up gearbox, and produced 12 kW of electrical power. Instead of using the power only as it was generated, Brush connected the dynamo to 408 batteries in the basement of his home. The electrical loads included 350 lights, two arc lights, and three electric motors. This was quite a feat for a first attempt at power generation from wind using nineteenth century technology.

In 1891, a Danish inventor by the name of Poul La Cour developed the first turbine to use four-bladed rotors with an old-fashioned airfoil shape. This design continued to advance and improve through World War I, resulting in a number of such systems across Denmark. These systems typically generated up to 25 kW of power. Industrialization again began to take its toll, however. Fossil-fuel plants producing large volumes of power continued to appear across the land.

Over the next 40 years, the electrical grid continued to expand. However, those that remained beyond the grid's reach commonly used wind turbines to generate local power. Another of America's wind pioneers, Marcellus M.L. Jacobs, installed his first wind generator with his brother Joe at their Montana ranch in 1922. Before long, they were organizing additional installations at the request of neighbors who saw the value. The Jacobs Wind Electric Co. continues in operation today as the oldest active wind turbine company in the United States. The company now provides turbines mostly for rural electric grids.

The state of Vermont claims ownership of the world's first megawatt-generating wind turbine connected to the grid. The Smith-Putnam 1.25 MW unit was commissioned in 1941. It used a twin-blade, 53 meter (175 feet) diameter rotor with speed controls to maintain 28 rpm.

Demand for electrical power in both homes and businesses continued to grow. Wind systems alone could not satisfy the needs. The use of wind power through the 1930s and 1940s was also impacted by government actions during the Great Depression. To stimulate rural economies especially, the electrical grid continued to expand through government investment.

Through the 1950s, power generation from wind continued to shrink in the United States. However, its advance persisted in Europe and other areas where wind and economic conditions favored its use.

German work in the field of wind power in the 1960s was led by Ulrich Hutter. He advanced lightweight plastic and fiberglass blade designs. His design focused on weight and avoiding damaging aerodynamic loads. The Danes had largely concerned themselves with structures that could withstand such loads.

The first major oil crisis in 1973 led to renewed worldwide interest in wind power. Wind farms again began to spread through the United States and abroad. Through the 1980s, attention turned from smaller (25 kW range) turbines to commercial turbines that could develop 50 kW or more

of power. Connecting them to the grid became possible. During the period 1981 through 1990, 17,000 wind turbines providing up to 350 kW each were built in California mountain passes (*Figure 5*). Consistent winds and government regulations favored their construction there. Not surprisingly, each national or world event that affects fossil-fuel price and availability also brings fresh interest in all renewable energy sources.

By the end of 2002, US wind turbines were providing roughly 4,600 MW of power. California and Texas were the sites of the most advancement. One year later, US capacity had grown to 6,300 MW. However, Europe's attention to wind remained more consistent, leading to the region's ownership of 70 percent of the world's wind power in 2003. The German wind industry led the European region with 14,000 MW of power production. It was enough to supply 3.5 percent of the country's power and employ 35,000 people at that time. During the same period, the Danes possessed the largest proportion of wind power. They were generating 20 percent of the their power and assembling nearly 40 percent of the turbines used worldwide.

74105-11_F05.EPS

Figure 5 Wind farm in California's San Gorgonio Pass.

On Site

The Work of a Previous Wind Energy Student

The inspiration for modern wind turbine designs came from a Danish engineer by the name of Johannes Juul. He was an early student of Poul La Cour's Wind Electrician Training Program. Juul designed and built the Gedser turbine in 1957, which used three blades on an upwind rotor and generated 200 kW of power. Juul is also credited with the design of the emergency blade tip brake; they are still used on turbines today.

2.2.0 The Wind Industry Today

Today's wind industry continues to expand dramatically across the globe in terms of both technical advancement and installation. Interest in renewable energy grows as fossil-fuel issues continue to emerge. The attention given to fossil fuels and the effects of their use has never been more consistent. Thus, the interest in developing wind power has followed suit.

Wind energy contributed over 166,000 MW of power globally by the end of 2010. US capacity represented over 36,000 MW of the global total. Germany and China were generating roughly 26,000 MW each.

By 2010, the United States was led by Texas in total wind power capacity. Adding the capacity of projects under construction in 2010 to the existing capacity will allow Texas to exceed 10,000 MW in power generated by wind. With one megawatt able to power roughly 300 homes, enough capacity will be on line to power 3,000,000 Texas households. The southeastern United States lags well behind the other regions, with scarcely any installations. This is due to the poor wind resources in the area. Although this trend will likely continue for the near future, technology continues to improve performance. Systems that can extract power from less wind may eventually open up calm territory to wind power investment.

Offshore projects (*Figure 6*), such as this one off the coast of Denmark, represent huge opportunities to capitalize on wind energy. This is because wind turbulence is reduced, and wind speed is higher and more consistent. The environment can also handle turbines of the largest size. Offshore installations allow turbines to be installed far enough away from populated areas where they are rarely seen from shore. The curvature of the Earth helps hide them from view. The first offshore wind farm serving the United States is located 16 miles off the coast of Nantucket, MA. The Cape Wind Project (*Figure 7*) will have 130 turbines and produce nearly 500 MW of power. A federal lease of 28 years was signed in October 2010. Although the construction costs are typically much higher offshore, other advantages make them attractive investments. Offshore installations are being discussed for the coastal areas of the southeastern United States as well. This may help that region to offset the lack of wind on land.

Thanks to the ability to communicate and share information globally, many organizations and government agencies have come to support wind energy. Efforts to create a cooperative and consistent industry environment are important. Many of these organizations are referred to in this text. An awareness of them would be essential to your future success in the industry. Now is a good time to become familiar with the industry players.

74105-11_F06.EPS

Figure 6 Offshore wind energy site.

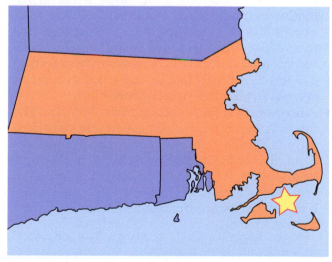

74105-11_F07.EPS

Figure 7 Cape Wind project location.

- *US Department of Energy (DOE)* – The DOE (www.energy.gov) has a broad mission. It includes energy security; nuclear security; scientific discovery and innovation; environmental responsibility; and effective energy management. The DOE's Wind and Water Program supports wind energy through partnerships to advance wind technology. The program also provides for collaboration with the electric power industry as a whole to use wind power on the grid without sacrificing reliability. The DOE has set a strategic goal of 20 percent national power generation from wind by 2030. A report released in September 2010 examines the feasibility and benefits of large scale, offshore wind power. The annual *Wind Technologies Market Report* can be found on its website and is a good read for anyone interested in the industry. The DOE's efforts benefit the entire US renewable energy industry. The DOE also provides support to the American Wind Energy Association (AWEA), discussed in more detail later in this section. Their primary goal is to ensure that the United States has an active role and a strong voice in the creation of international standards. A large part of the DOE's support is provided through the National Renewable Energy Laboratory.

- *The National Renewable Energy Laboratory (NREL)* – The NREL (www.nrel.gov) conducts research in wind energy systems at its National Wind Technology Center (NWTC). The center is located south of Boulder, CO. A variety of testing is conducted there and at various satellite locations. Newer facilities enable the testing of blades up to 100 meters long. Testing and research conducted at the NWTC has contributed to the success of many large wind farms. Funding comes in part from the DOE. The mission of NWTC is to work with private industry to advance wind technology and accelerate its commercial use. NWTC's research and testing has led to a great deal of the wind industry's success in effective power generation.

- *American Wind Energy Association (AWEA)* – The mission statement for AWEA (www.awea.org) is quite simple: to promote wind power growth through advocacy, communication, and education. AWEA has the responsibility to develop and publish standards for the US wind industry. Standards are first reviewed and adopted by the American National Standards Institute (ANSI). ANSI recognizes AWEA as an Accredited Standards Developer for the US wind industry. With that responsibility, AWEA also participates in standards development globally. It cooperates with an array of other standards-producing organizations, such as the American Society of Mechanical Engineers (ASME), the American Society for Testing Materials (ASTM) International, and the Institute of Electrical and Electronic Engineers (IEEE). Internationally, the organization is an active participant in the International Energy Agency (IEA); the International Electrotechnical Committee (IEC); and the International Standards Organization (ISO). These groups are responsible for the majority of accepted standards in many US industries.

- *World Wind Energy Association (WWEA)* – WWEA (www.wwindea.org) is an international non-profit organization that blends the efforts of over 90 member countries to promote the use of wind energy globally. One of the most important benefits of the group is the world-wide sharing of technology and research data.

There are many other organizations making important contributions to the wind industry. However, the efforts of these organizations will likely have the greatest impact on the future of the industry.

2.3.0 The Future of Wind Power

The DOE has set a goal of providing 20 percent of US electrical needs from wind by 2030. The goal is part of the DOE's Wind Powering America (WPA) program. In 2008, a report was published that assesses the feasibility of reaching that goal. The report, *20% Wind Energy by 2030: Increasing Wind Energy's Contribution to US Electricity Supply*, includes contributions from the DOE and its national laboratories. The wind industry, electric utilities, and other groups also assisted. The report examines the challenges of producing 300 GW of wind power, representing 20 percent of projected US energy needs, by the target year. Some of its more important findings include the following:

- An enhanced power transmission system is needed, along with a more efficient permitting and siting process.
- The number of turbine installations must increase from 2,000 units per year in 2006 to roughly 7,000 units per year by 2017.
- The integration of 20 percent wind power with the power grid can be done for less than 5 cents per kWh.
- The goal is not endangered by the availability of raw materials needed for turbine and site construction.

In 2000, when the DOE's wind efforts began, only 2,500 MW of wind energy capacity was on line. Only four states had more than 100 MW of capacity. By 2009, 22 states had over 100 MW of capacity on line. It is expected that 30 states will surpass this mark by 2020. Industry analysts see this as a very positive gain, because the first 100 MW of turbines in a state are the most difficult to achieve. Both residents and state governments are somewhat resistant until the effects are fully realized. They must first understand the importance and relevance of a renewable energy future, and that process takes time. Once the idea is embraced and the first few installations are done, the stage is set to proceed at a faster pace.

In an effort to educate the public about the benefits of wind energy, the WPA program has also established learning opportunities in schools across the United States. Wind Application Centers (WACs) have been established at a number of universities, which now offer engineering classes in wind power technology. K-12 schools are also participating. Three to five host schools per state have been sought, where wind energy lessons are introduced into the science programs. It is hoped that the program leads to the installation of small wind turbines at the host schools. The turbines provide students with a working example to deepen their understanding.

The science of wind power continues to expand. The development process continues for every required component. The development of an advanced rotor blade with gently curved tips (*Figure 8*) is expected to increase energy capture by 9 percent and function well at lower wind speeds. Generators that develop power at lower speeds and improved gearboxes are tested constantly. Gearboxes, which increase rotor speed to a level high enough for power generation, represent the bulk of maintenance and repair costs. As a result, several manufacturers are developing turbines with generators able to develop power at low speeds, eliminating the gearbox. By the time 2030 arrives, the needed 300 GW of wind power may be possible with fewer turbines as the direct result of research and development.

2.4.0 Wind Power Career Opportunities

According to studies at the University of California, the wind industry is more labor intensive than other energy technologies. Job opportunities come in manufacturing, scientific, engineering, and service roles. Domestic jobs will grow from consistent investment in wind power.

Many manufacturers prefer to invest in support and manufacturing facilities close to major wind sites. This is partly the result of the staggering costs associated with transportation of the large

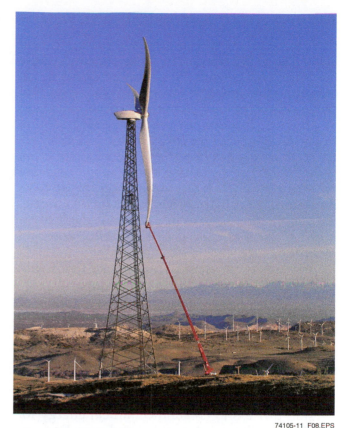

74105-11_F08.EPS

Figure 8 Sweep Twist Adaptive Rotor (STAR) turbine blade.

components. Many wind turbines and related components are currently manufactured outside of the United States. However, domestic production is increasing. Sustaining it will depend upon consistent renewable energy goals being developed by federal and state governments. Complex permitting and licensing processes are a concern to US manufacturers, since they affect the demand for new turbines.

According to a report from the AWEA, about 85,000 workers were employed in the wind industry by 2008. If the goal of 20 percent wind energy for the United States by 2030 is met, the industry could support 500,000 jobs. Roughly 180,000 of those jobs would be directly related to the industry through construction, operations, and manufacturing processes. Between 2020 and 2030, pursuing the goal would support 100,000 jobs in related industries. These jobs include professions and trades such as accounting, law, ironworking, and component manufacturing.

Due to the growth of wind sites, a shortage of qualified workers already exists in the industry. The service and installation of wind turbines require a combination of skills and qualifications. Both tasks require the ability to work at great heights. Extensive travel is also required of some workers, especially those who work directly for manufacturers providing warranty or startup services.

3.0.0 A STUDY IN WIND ENERGY

Before discussing the energy in wind and how it is harnessed, it is important to understand where the wind comes from. Wind can be called a side effect of solar energy, created by the impact of the sun heating our planet at various rates and at different locations. Further, water and land features offer different colors and levels of reflectance, affecting heat absorbency. As the air above is heated, it begins to rise and create a negative pressure beneath it (*Figure 9*). Nature dislikes a vacuum, so cooler air rushes in below, replacing the air that has risen. This movement of air masses is the wind, at varying rates of flow and direction.

Once a mass of air is in motion, it carries with it the energy that created that motion. This is **kinetic energy**, the energy contained in a mass or body caused by its motion. A car moving at any speed still contains the kinetic energy required to reach that speed, even if the engine is turned off. Through wind turbines, the kinetic energy of the wind is converted into both mechanical and electrical energy. Humans have applied this conversion process to create mechanical energy for pushing sailboats, pumping, and grinding operations. The focus, of course, is how it is used to generate electrical power.

3.1.0 The Power of the Wind

It is simple enough to say that the amount of energy in the wind is a mathematical relationship between its velocity (speed) and mass. But there is a lot to learn about both factors to better understand the result of the calculations.

Wind speed is the first factor to consider in the energy that can be extracted. It changes constantly, as we all know, sometimes moment by moment. The speed is affected by many factors, including time of day (or night), since the sun is the primary generator of wind. Further, since the sun changes its position in the sky to some degree on a daily basis, its impact on a given point on Earth's surface changes too. Global and seasonal weather patterns also affect wind speed and its consistency.

Although wind speed is intermittent and unreliable by nature, some means of discussing and documenting its speed at a given location must be identified. Using an average speed over some period is the generally accepted method. The average wind speed over the course of one year is an accepted standard. It should be noted, however, that these averages could change significantly year to year. Wind speed is important because the faster the wind, the more energy it contains.

The second factor in determining wind energy is its mass. At any given speed, there is more kinetic energy stored within a heavy object than a lighter one. Would you prefer being hit by a marble at 40 kilometers per hour (kmh), roughly equal to 25 mph, or by a bowling bowl at the same speed? Although air is very light relative to most other things, it does have measureable mass. A cubic meter of water, due to its mass, can store more energy than a cubic meter of air, but the wind has volume and speed on its side to help make up for its low mass.

The mass (m) of air is defined as the product of the air density (represented by ρ, the Greek symbol rho) and the volume. To calculate the volume of air, multiply the wind velocity (V) by the area (A) it is passing through for some chosen period of time (t). See *Figure 10*. The area will be equal to the size of the circle created by the turbine rotor as it turns. This concept will be explored in much greater detail in this module.

Mathematically, the complete mass equation would be:

$$\text{Mass} = \text{density} \times (\text{area} \times \text{velocity} \times \text{time})$$

or

$$m = \rho A V t$$

Once the mass is known, the kinetic energy can be determined with an additional calculation that relates the wind speed (V) to the mass. The mass calculation can be combined with the one for energy so the math can be completed in one equation:

$$\text{Wind energy} =$$
$$\tfrac{1}{2} \times \text{density} \times (\text{area} \times \text{velocity} \times \text{time}) \times \text{velocity}^2$$

or

$$\text{Wind energy} = \tfrac{1}{2}\rho A t V^3$$

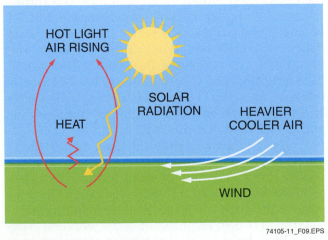

HOT LIGHT
AIR RISING

SOLAR
RADIATION

HEAVIER
COOLER AIR

HEAT

WIND

74105-11_F09.EPS

Figure 9 Wind generation.

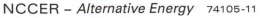

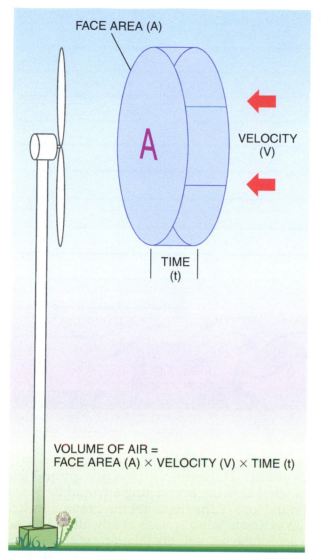

FACE AREA (A)

A

VELOCITY
(V)

TIME
(t)

VOLUME OF AIR =
FACE AREA (A) × VELOCITY (V) × TIME (t)

74105-11_F10.EPS

Figure 10 Air volume calculation illustrated.

The units of measure for wind energy are discussed in a later section.

It is important to keep in mind that increasing the speed or the mass of the wind increases the available energy. But increases in wind speed create a far greater energy increase than an increase in mass, due to the cubic relationship of velocity (V^3) in the equation.

The equations may look complex, but understanding the concept is most important. Speed and mass are the two primary factors in wind energy. The total energy in wind that can be captured is quantified by relating its speed and mass, the area it passes through (rotor diameter), and the amount of time that passes.

3.2.0 More About Wind Velocity

For any size of rotor, the wind speed has the most impact on power produced. As mentioned earlier,

this is the result of the cubic relationship of speed in the power equation. We can look at some simple statements about the impact of wind speed on the power a turbine can capture:

- Consider a wind speed of 4.5 meters/second (10+ mph), and an increase of 10 percent up to 4.95 m/sec (11.1 mph). This small change results in a 33 percent increase in available power.
- Increasing the speed by 20 percent, to 5.4 m/sec (12.1 mph), increases the power by 73 percent.
- Doubling the speed from 4.5 m/sec to 9 m/sec (20 mph) increases the available power 8 times.

These are important facts to remember as you continue through the program. For our purposes, the math is not as important as grasping the concepts.

Calculations for the effect of wind speed on available power must be based on either a single speed at any given moment, or an average speed over some period of time. Annual averages are often used in the industry for many discussions. However, the time base of the average used must be carefully considered. Remember that an average is based on many values that are either above or below the average. To predict available power more accurately, engineers and designers often must make many individual calculations and add them together. For example, power may be calculated at a site based on the number of hours per year for one speed, and then factoring in the number of hours another speed is maintained, and so on. In the end, the sum of the results would more accurately represent the power available than a single calculation using an annual average.

3.3.0 Wind Velocity and Height

The wind at ground level may seem intense at times, but it pales in comparison to the wind speed (and thus the power available) at higher levels. Further, it doesn't take a significant height increase to find far more wind. Just as friction in piping impedes water flow, wind movement across the terrain is affected by friction (*Figure 11*). Trees, buildings, and hills are just a few of the many obstructions that slow the wind by friction. Of course, some obstructions have a greater impact than others do. The wind blowing across a lake, for example, encounters less friction than it does when blowing across a forest of trees. This is the reason that offshore locations, in spite of the cost of construction and maintenance, offer superb wind resources.

As a rule, then, wind velocity is greater at higher levels above the earth than at its surface. The obstacles on Earth create varying degrees of friction, causing the air to tumble or move er-

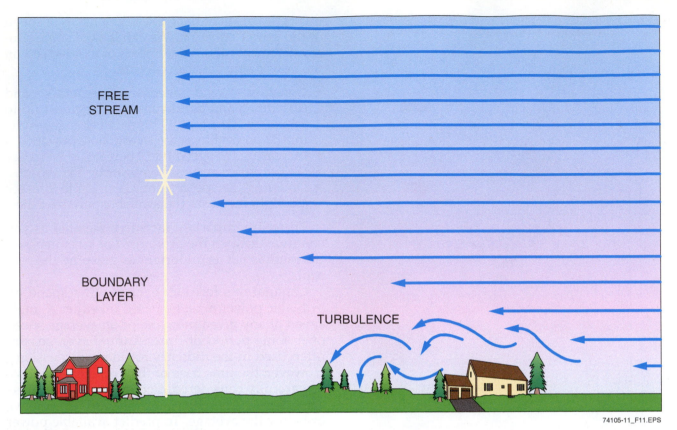

Figure 11 Effect on wind from ground obstructions.

ratically. The smoother flow of air above all the obstructions is therefore faster. Of course, there will always be exceptions to the rule. Different terrains and obstacles also have different effects. Over mountains or hills, wind speed tends to increase more dramatically with increasing height. Radical changes in terrain, such as tall buildings or mountain cliffs, can affect air movement well above the height of the obstruction itself.

The objective is to place a wind turbine in the best air stream available. This not only applies to a region, but also to the local level. Although it is widely accepted that Mt. Washington in New Hampshire is the windiest location in the United States, it does not necessarily mean that a wind turbine installation anywhere in that area is a good idea. At the local level, velocity and power production may be best at 100 meters above ground in one spot, but even better and more consistent at another spot at 80 meters.

This phenomenon of wind speed varying at altitude due to friction on Earth's surface is known as **wind shear**. This should not be confused with wind shear as it applies to the aviation industry and flight. It should be noted that the basic theory of wind shear is true primarily for heights above the **effective ground level**. As the wind blows across a smooth lake, the effective ground level would be the surface of the water. Wind shear

decreases above this level, allowing the wind to move faster. However, the effective ground level of a wind blowing over a jungle would be considered the tops of the trees, not the ground itself.

To estimate wind speeds at any altitude based on actual measurements taken near the ground, we must be able to express wind shear mathematically. The value used is known as the wind shear exponent, which itself can be calculated based on values assigned to surface roughness. *Table 1* provides a list of different terrains, along with the surface roughness factor assigned and the corresponding wind shear exponent.

The table shows how wind shear exponents increase as the terrain becomes rougher, offering more resistance to air flow. As a result, a more dramatic rate of increase in wind speed above those surfaces occurs as altitude rises. On average, or when the wind shear exponent is not identified in a discussion, it is generally accepted to be that of short or mown grass areas, or an exponent of 0.143 to be precise. Since 0.143 represents roughly ⅐ of 1, the wind shear effect is known as the ⅐ power law. The ⅐ power law applies to many sites across the United States, allowing engineers to estimate wind velocity at heights based on a common wind shear exponent. Remember, though, that not all such rules of thumb are perfect and apply to every case.

Table 1 Wind Shear Exponents for Various Surfaces

Type of Terrain	Surface Roughness Length (m)	Wind Shear Exponent
Ice	0.00001	0.07
Snow on flat terrain/calm water	0.0001	0.09
Snow-covered low crops	0.002	0.12
Short or mown grass	0.0007	0.14
Crops or prairies with taller grasses	0.05	0.19
Scattered tall shrubs and trees	0.15	0.24
Above with a few buildings added	0.3	0.29
Suburban areas with homes, etc.	0.4	0.31
Wooded areas/forests	1	0.43

For locations where the $\frac{1}{7}$ power law applies, other rules of thumb can be calculated based on it. For instance, at some locations, doubling tower height increases wind speed about 10 percent, while increasing height by a factor of 5 (quintupling) increases speed roughly 25 percent.

3.4.0 Wind Data Acquisition and Use

There are clearly ways to estimate and predict wind speed and the available power. However, there is nothing better than actual, real-world data recorded over a long period of time. The financial investment of a wind energy site is very high. Proceeding without data measured using an **anemometer** at or near the exact location of the rotor center would be foolhardy.

3.4.1 Met Towers

The meteorological tower, or met tower, is an important aid in determining the wind properties at a given location. Met towers in the 30- to 50-meter range can be set up in a reasonably short period. The data acquired at several locations in a local area can assist in creating trend maps of the area. Permanent met towers are sometimes built near wind turbine farms as a back-up source of information. Turbines also have instruments mounted on them. Where changes in surface terrain are significant, more sites and data are likely needed to make accurate predictions. Along with speed,

the wind direction and temperature are usually recorded. Data can be stored on site through a microprocessor and memory card, or transmitted via cellular technology to a central location. Not only is site-specific data needed to select the best site for the wind turbine, such data is also important to potential investors.

3.4.2 Wind Maps and Charts

In the absence of site-specific data, which comes at a cost, a means of quickly evaluating wind conditions on a larger scale is needed for a number of purposes. The National Climatic Data Center (NCDC) headquartered in Asheville, NC contains weather data collected from a variety of federal resources, the most notable of these being the National Weather Service (NWS) and the Federal Aviation Administration (FAA). More than 30 years of information has been amassed for some locations in the United States. It is highly useful data, but remember that wind properties can vary dramatically even at the local level.

Figure 12 is a map provided by the NREL of the average annual wind resource for the continental United States. An incredible array of such maps is available for review, including those showing the level of certainty for estimated winds in various areas. These same maps can also be found for one season of the year. If one can think of a way to report and correlate wind data, it likely exists as a map or tabular report from the NREL.

Collecting Wind Data

Basic wind data collection requires only a few instruments. Cup anemometers are the most common. A wind directional sensor is mounted on the same tower on the left side. The propeller anemometer style is also used in some applications.

A wind data logger is needed to record the desired information over a period of time. The model shown here, which is also available with a solar power kit to eliminate utility connection or batteries, offers the following features:

- The ability to record wind velocity, gusts, and wind direction from multiple anemometers simultaneously
- Logging of temperature
- A real time clock
- Data logging interval programming of 10 seconds to 16.6 hours in one-second intervals of choice
- An RS-232 port for connection and data download to a computer
- The ability to accommodate many other sensor types, including relative humidity, rainfall, and light intensity

CUP ANEMOMETER

74105-11_SA01.EPS

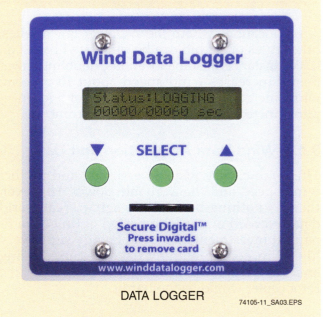

DATA LOGGER

74105-11_SA03.EPS

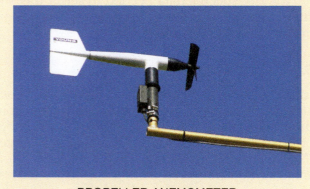

PROPELLER ANEMOMETER

74105-11_SA02.EPS

Other entities have also done estimates and compiled data from both federal and private sources to gain a more accurate picture of wind properties. This is especially true for areas that are regarded as good sites for wind power generation.

A variety of wind information and wind maps can be accessed from the following sources for further study:

- *National Climatic Data Center* – www.ncdc.noaa. gov
- National Renewable Energy Laboratory (NREL) – www.nrel.gov and www.nrel.gov/rredc
- *US Department of Energy, Wind Powering America Program* – www.windpoweringamerica.gov

Another useful tool derived from years of recorded data is the wind rose (*Figure 13*). This circular graph depicts the frequency and speed that winds blow from a given direction, reported as some percentage of time. Each concentric circle represents a different frequency of time, starting from zero at the center to higher percentages at the outer circles.

4.0.0 INTERCEPTING WIND ENERGY

In the previous section, the face area of the air passing by as a function of the wind energy equation was mentioned several times. The area we are concerned with is called the swept area—the area that the turbine rotor blades pass through.

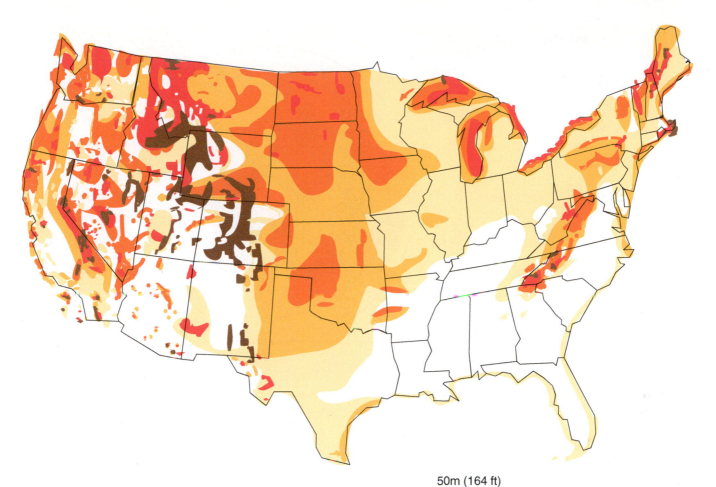

POWER CLASS	WIND POWER (W/m²)	SPEED* (m/s)
1	<200	<5.6
2	200-300	5.6-6.4
3	300-400	6.4-7.0
4	400-500	7.0-7.5
5	500-600	7.5-8.0
6	600-800	8.0-8.5
7	>800	>8.5

50m (164 ft)

*Equivalent wind speed at sea level for a Rayleigh distribution.

74105-11_F12.EPS

Figure 12 Average annual wind resource estimated availability for the contiguous United States.

Greater energy is available to be intercepted over a larger swept area. Thus bigger is better, as long as the entire rotor remains active and is not somehow blocked.

A common horizontal axis (propeller-style) turbine rotor sweeps a circle as it turns. The swept area of a circle is easily calculated as follows:

$$\text{Area} = \text{pi} \ (3.1416) \times \text{radius}^2$$

or

$$A = \pi r^2$$

Although this is a relatively simple equation, it is very easy to confuse radius with diameter since wind turbine discussions may use both terms.

Again, more area is better. As blade length is increased, the proportional increase in swept area

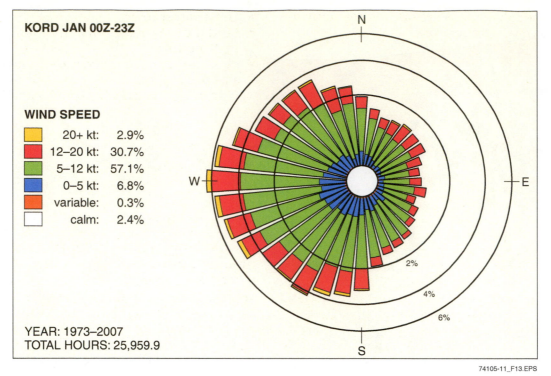

KORD JAN 00Z-23Z

WIND SPEED

■	20+ kt:	2.9%
■	12–20 kt:	30.7%
■	5–12 kt:	57.1%
■	0–5 kt:	6.8%
■	variable:	0.3%
□	calm:	2.4%

YEAR: 1973–2007
TOTAL HOURS: 25,959.9

74105-11_F13.EPS

Figure 13 Wind rose.

is dramatic. Doubling the length of the blade (effectively doubling the diameter) quadruples the swept area. With an understanding of the principle of swept area, the power that can be intercepted can be analyzed.

It was previously mentioned that average wind speeds often misrepresent the true characteristics of a site. For example, at any site, a given wind speed may occur 15 percent of the time; at another it occurs 6.2 percent of the time, and so on. As a result, it is often far more accurate to sum the amount of power available based on many calculations on a range of speeds than a single calculation using a time-based average.

To quantify the amount of power available, the term **power density** is used and power in units of watts per square meter or w/m² are measured. In English units, it is expressed in watts per square foot, or watts/ft². Power density is represented by P/A in the equation. The calculation takes speed as well as air density (to quantify the air's mass) into consideration.

Assume a site's wind has been monitored and the calculations lead to a power density of 60 w/m². From here, the power available for an entire year can be calculated (expressed as E/A) through multiplying the power density by the hours in a year (8,760) and changing the watts to kilowatt/hours, as follows:

$$E/A = P/A \times \text{hours of chosen period}$$
$$\times (1 \text{ kW}/1,000 \text{ watts})$$
$$E/A = P/A \times 8,760 \text{ hours} \times (^1/_{1,000})$$
$$E/A = 60 \times 8,760 \times (^1/_{1,000})$$
$$E/A = 525,600 \div 1,000$$
$$E/A = 525.6 \text{ kWh/year/m}^2$$

Note that these values are the amount of power available per unit of area; in this case, square meters. This can now be related to the swept area of the rotor. For the example site that provides 526 kWh/year/m², assume a rotor with a diameter of 40 meters (a radius of 20 meters). The swept area of a spinning propeller-style rotor is calculated the same as any circle:

$$A = \pi r^2$$
$$A = 3.1416 \times 20 \text{ meters}^2$$
$$A = 3.1416 \times 400$$
$$A = 1,256.64 \text{ m}^2$$

Since our site provides 526 kWh/year/m², simply multiply this number by the area of the rotor to determine the amount of power the rotor will see or intercept over the course of a year:

Total E/A = 526 kWh/year/m² × 1,256.64 m²
Total E/A = 660,993 kWh/year intercepted by the 40-meter rotor at this set of conditions

Again, the calculations are offered here only to help demonstrate the concepts.

4.1.0 The Betz Limit

In a perfect world, people would be able to transfer all of the available 660,000+ kWh through the rotor and make use of it. But as you might expect, it isn't possible. There are losses associated with the process. In order to extract all of the energy, the wind would have to come to a complete stop since it would be drained of all kinetic energy. But that doesn't happen. After making contact with the rotor blade, it works its way around it and continues on its way.

A German scientist named Albert Betz is credited for developing what is known as the **Betz limit**. This is the theoretical limit of how much of the wind's power a rotor can capture. Per Betz's calculations, only 59.3 percent of the power is truly available. Again, this is the theoretical limit for all rotors, regardless of any claims that it has been exceeded. It should not be confused with how much is actually captured by every rotor. In reality, rotors rarely approach even this level of efficiency. It is the target value for perfect efficiency from a rotor. Keep in mind, though, that designing the perfect rotor may not be so effective in the end. The perfect wind turbine provides maximum electrical power for the lowest cost in the real world. A turbine equipped with a perfect rotor may not necessarily accomplish that.

To apply the Betz limit, multiply the total E/A by the value of 59.3 percent, or 0.593. Considering the result of the math in the previous section, the limit of power the turbine can really intercept, then, with a perfect 40-meter rotor at this site, would be 391,969 kWh/year.

Large rotors that can extract 40 percent or more of the wind's power have been developed. Smaller rotors capture far less than this—some as little as 12 percent. In general, rotors in the real world typically capture from 12 percent to 40 percent of the available power. Smaller rotors are the least effective at power capture.

It should be noted that even though a rotor may capture 40 percent of the available power, not all of that would actually be transformed into electrical power. There are still many other losses to consider, including bearing friction. But an earlier concept remains true here – the rotor is the limiting factor when it comes to power capture from the wind. If the calculations indicate that the energy produced with a certain wind turbine is insufficient, then there are three simple ways to increase it:

- Get a bigger rotor.
- Feed the rotor more wind; for example, by finding a new site.
- Use a turbine that has more efficient parts (such as a better generator), which will not often provide much increase in power generation.

5.0.0 WIND TURBINES

Wind turbines come in many shapes and sizes. There are tiny propeller toys and wacky contraptions in the backyards of home inventors. Many different configurations are attempted. Some are successful; others are not. What constitutes a successful design may be a matter of opinion. Many unique approaches have resulted in successful power generation, but not necessarily in enough volume or in a cost-effective manner. This section examines at the basic designs of common commercial and utility-scale wind turbines.

5.1.0 Basic Designs

Turbine designs typically fall into one of two broad categories: **horizontal-axis wind turbines (HAWTs)** or **vertical-axis wind turbines (VAWTs)**.

VAWTs (*Figure 14*) have two major advantages in their design. They can accept wind power from any direction (omnidirectional). In addition, the bulk of their major parts, such as gears and generators, are located at ground level for easier access and service. A number of shapes and designs have been attempted over the years. Many of them are inspired by designs from a French inventor named D. G. M. Darrieus in the 1920s. One of the more popular Darrieus-inspired designs is the phi, or eggbeater, configuration (*Figure 15*).

Many large VAWT designs have suffered from structural or performance issues; they have achieved limited success in commercial and utility-scale applications. However, variations of the H-pattern VAWT, as well as other small vertical designs, have proven reliable. They can be fairly inexpensive to purchase and install. They can be used commercially when the environment allows for their installation in significant numbers.

HAWTs, on the other hand, have enjoyed great success in all market sectors and they are the primary focus today (*Figure 16*). They do require turning of the rotor into the wind and the major parts are located high in the air. However, these disadvantages have been overcome and their reliability and performance is generally far better than VAWTs in large applications.

Figure 14 H-pattern VAWT.

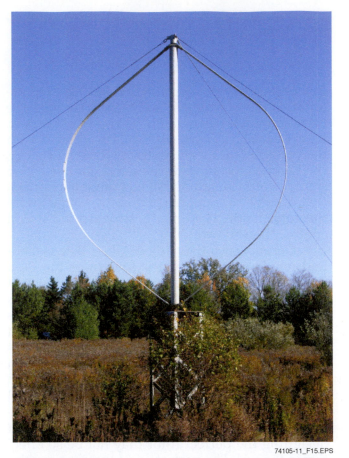

Figure 15 Phi, or eggbeater style, VAWT.

5.2.0 HAWT Systems

HAWTs vary somewhat in their design. Although most position the blade upwind (*Figure 17*), some smaller designs use a downwind rotor. Most downwind models use a very simple means of positioning the rotor into the wind, such as a tail vane. Others utilize a self-aligning blade design. Smaller upwind turbines also may use a tail vane (*Figure 18*). Larger, heavier turbines cannot be turned into the wind reliably by a simple tail vane and must be mechanically moved to the proper position.

A complete commercial turbine system consists of a variety of parts and subsystems. The complete system includes the turbine itself, its foundation and supporting tower, and a **supervisory control and data acquisition (SCADA)** system. A great deal of wiring and electrical parts are also part of the system.

Figure 19 shows the major parts of the turbine. The following is a list and brief description of the individual components shown:

Figure 16 Typical HAWT.

- *Pitch* – Refers to a twisting motion of the entire blade, changing how the wind impacts it. The arrow indicates the rotation of each blade as the pitch is changed to accommodate the condition.
- *Low-speed shaft* – Directly connects to the rotor itself, turning at the same rpm and transferring wind power to the gearbox.

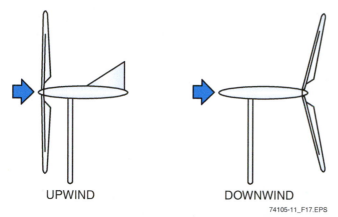

UPWIND   DOWNWIND

74105-11_F17.EPS

Figure 17 Upwind and downwind rotors.

- *Gearbox* – Houses a set of gears and accepts the power input from the low-speed shaft. Through gearing, the rpm is increased to a usable value for the generator to create power.
- *Generator* – Develops electrical power from mechanical energy input through the high-speed shaft.
- *Controller* – Takes in information from the anemometer, wind vane, and other sensors, and controls the pitch and yaw of the turbine. The internal controller may be interfaced with, or replaced by, a SCADA system.
- *Anemometer* – Measures wind speed.
- *Wind vane* – Monitors wind direction.
- *Nacelle* – Streamlined housing or enclosure which contains the major working components of the turbine system at the top of the tower.
- *High-speed shaft* – Transfers the higher rpm output of the gearbox to the generator.
- *Yaw motor* – One or more provide the power to pivot the nacelle assembly.
- *Yaw drive* – The gearing that turns the turbine assembly on its vertical axis.
- *Brake* – Equipped to slow the rotor and bring it to a stop for maintenance, or in the event of an emergency such as a failure of yaw and/or pitch controls.
- *Tower* – Supports the nacelle and provides a means of access from the ground.
- *Blades* – Intercept the wind and turn wind energy into mechanical rotation.
- *Rotor* – Provides a mounting point for the blades and connects to the main low-speed shaft.

5.2.1 Blades

Another significant HAWT design variance relates to the number of blades. Single-blade turbines certainly work, and proponents believe them to be cost-effective performers. However, efficient energy capture, strength problems, and aerodynamic issues hinder their effectiveness.

74105-11_F18.EPS

Figure 18 Small HAWT with tail vane.

Since a balancing counterweight (which weighs as much as the single blade) is required, some losses due to useless weight in motion occur. Two-blade and three-blade systems offer obvious balance advantages, as does the multi-blade rotor (*Figure 20*). The vast majority of commercial turbines use three-blade rotors, as they have been found to operate more smoothly, quietly, and effectively than all other models.

HAWT blade sizes vary depending upon the use and desired output. The rotor diameter and swept area of the rotor is the main controllable factor in wind capture. Typical modern large wind turbines have blade lengths of 20 to 45 m (65 to 150 feet) and produce between 500 kW and 2 MW of power.

A number of turbines, as small as 50 watts, are available to provide minimal power for small uses like boating and camping. Since power storage in batteries is often important in smaller applications, they typically produce DC power rather than AC. Turbines of this size are at the entry level of the growing small wind market. This segment of the market continues to grow and the United States leads the world in their production.

New commercial turbines continue to increase in size and power production as technology provides improved materials and engineering. The world's largest unit presently in service is the German-built Enercon E126 (*Figure 21*) with 24 units on line as of 2010. With a hub height of 135 meters (443 feet) and a blade length of 63 m (206+ feet), this model can generate up to 7 MW of power. It is difficult to comprehend its enormous size in the photo. Research and engineering are presently

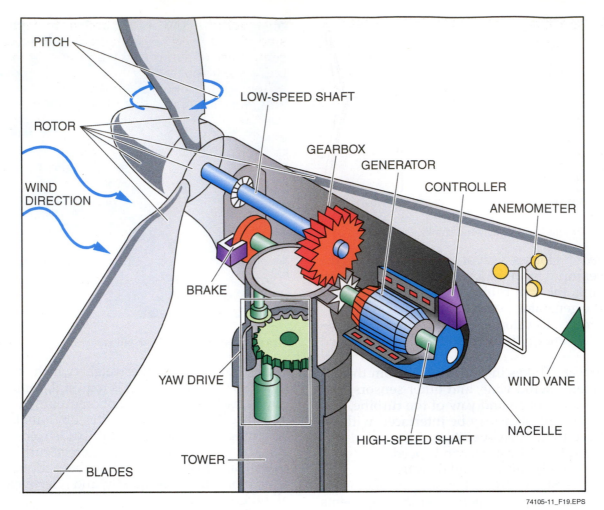

PITCH

ROTOR

WIND DIRECTION

LOW-SPEED SHAFT

GEARBOX

GENERATOR

CONTROLLER

ANEMOMETER

BRAKE

YAW DRIVE

HIGH-SPEED SHAFT

WIND VANE

NACELLE

BLADES

TOWER

74105-11_F19.EPS

Figure 19 Basic turbine system components.

under way by several groups to produce a 10-MW unit. Although turbines of this size may be built on land for testing, they are best used offshore because of their huge size. *Figure 22* provides a visual representation of the actual size of various turbines.

74105-11_F20.EPS

Figure 20 Multi-blade turbine rotor.

Many materials have been used in blade construction over the years. New materials are being constantly developed. Wood is still used on some small turbines, and wood composites have been used for blades up to roughly 43 meters (141 feet). Steel and aluminum have both been used. Steel is strong but heavy, while aluminum is light but subject to severe damage resulting from metal fatigue. Both have also been known to contribute to radio and TV signal interference.

Fiberglass is presently favored for most large turbine sizes due to its combination of low weight, strength, and cost. Carbon fiber is also used in some applications due to its desirable strength and rigidity, but at a much higher cost. A typical utility-scale turbine blade is shown in *Figure 23*. This photo helps show the actual size of wind turbine components. Transporting them to remote areas is challenging. Note that this blade is roughly half the length of the Enercon E126's blade.

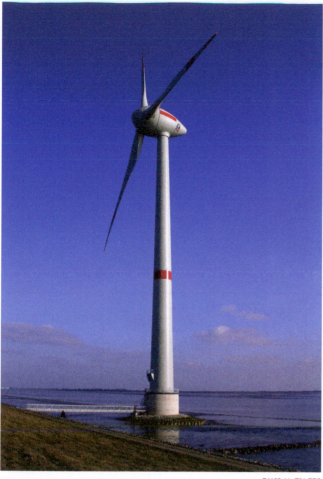

Figure 21 The Enercon E126 turbine.

5.2.2 Towers

The tower supports the main turbine assembly and provides a route for power transmission to the ground. It also provides maintenance personnel with some means of accessing the turbine. Of course, tower design and construction differ dramatically depending upon the size and height of the turbine.

Height is a significant factor in reaching strong winds. Taller towers place the turbine well above obstructions and allow access to faster and smoother air. As height increases, however, the complexity of designing and building a tower increases exponentially.

Towers fall into two broad categories: guyed and freestanding. Guyed towers (*Figure 24*) are primarily limited to use with smaller turbines and are more economical than their freestanding counterparts to build on this scale. They are impractical for very large turbines. Since the guy wires are responsible for maintaining the tower's top position, the weight-supporting portion of the structure is often consistent in size from top to bottom. The foundation is also minimal in size and mass.

Freestanding towers are further broken down into two categories: lattice towers and tubular towers. Lattice towers (*Figure 25*) are sometimes called truss towers. They may also use guy wires in smaller applications.

Lattice towers, depending upon their height, can be shipped in sections for site assembly or built totally on site. Heavy-duty hinges at the base can allow the assembly to be done on the ground, and then the tower can be pulled to an upright position. Opponents of lattice towers cite their unattractive appearance as a reason to avoid their use.

Tubular towers (*Figure 26*) are primarily used for large turbines. They are also referred to as monopole towers. Although they are more expensive to build, the wind industry in general prefers the tubular style for several reasons. They offer better appearance, security, improved personnel protection, and turbine access. The tallest towers may be equipped with lifts to prevent service personnel from having to make the daunting climb to the top.

Small tubular towers are built of steel pipe or tubing, concrete, or fiberglass. Like some lattice towers, simple pole towers can be hinged at their base to allow for easy erection. Multiple guy wires are then attached to maintain their posi-

---

On Site

Carbon Fiber-Reinforced Plastic (CFRP) Blades

Blade failures due to the unusual stresses they must endure have taken their toll on the industry. A batch of blades that must be recalled represents a huge financial burden and damages a company's reputation. In the quest for long-lasting blades, lighter weight would be a bonus as well. Reducing blade weight has a dramatic weight-saving effect throughout the rest of the wind turbine system.

Although quite expensive to date, one material offers three times the stiffness and improved fatigue properties over fiberglass and plastic. CFRP offers potentially thinner and stiffer blades. Their reduced mass makes them easier to handle and transport. This material is also the primary material of fuselage construction for the Boeing 787 Dreamliner.

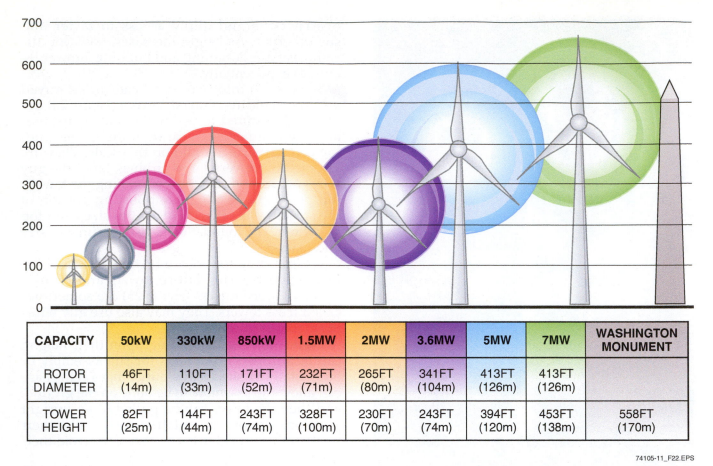

CAPACITY	50kW	330kW	850kW	1.5MW	2MW	3.6MW	5MW	7MW	WASHINGTON MONUMENT
ROTOR DIAMETER	46FT (14m)	110FT (33m)	171FT (52m)	232FT (71m)	265FT (80m)	341FT (104m)	413FT (126m)	413FT (126m)	
TOWER HEIGHT	82FT (25m)	144FT (44m)	243FT (74m)	328FT (100m)	230FT (70m)	243FT (74m)	394FT (120m)	453FT (138m)	558FT (170m)

74105-11_F22.EPS

Figure 22 Wind turbine sizes.

tion. As size and height increase, guy wires are eliminated and the tower is tapered to allow the base to be larger than the top for improved stability. Tubular steel towers can be made in sections and shipped to the job site for assembly (*Figure 27*).

Larger tubular towers are made of steel, or a combination of both concrete and steel. Since the tower base needs mass for stability, concrete may be better suited for the lower portions of a tower, while steel has the advantage in the slender upper sections.

Large tubular towers offer the advantage of providing safe access to the turbine for service (*Figure 28*). It is not difficult to understand how this type of access is far more attractive than the access a lattice tower provides. Lattice towers are especially hard to climb in extremely cold, windy, or wet climates.

Offshore wind installations offer great advantages, but they also offer unique challenges. Existing shallow water sites are little more than typical land-based towers that have been modified for better corrosion control and have marine-class

74105_F23.EPS

Figure 23 Typical utility-scale HAWT turbine blade.

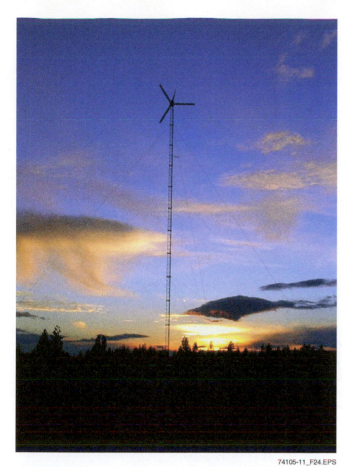

Figure 24 Guyed tower.

electrical systems. They are then anchored to the seabed. Placing turbines in deeper water increases the complexity of the tower design, since larger and more powerful turbine systems are desired to take full advantage of the high winds and avoid coastal traffic. An underwater guy system may be fine for some depths and tower sizes. However, deep sea installations will likely need to take advantage of the floating technology developed for oil rigs and drilling platforms. Obviously, the destructive force of hurricanes must be considered regardless of the installation approach.

5.2.3 Nacelles

The nacelle (*nuh-sell*) houses all of the wind turbine's working components and moving parts. The nacelle must be somewhat aerodynamic to avoid disrupting airflow as the wind passes through the rotor. If not, it can create undesirable backpressure. It must also have the structural strength to withstand all types of foul weather.

Figure 29 shows the internal layout of a nacelle from one manufacturer. Nacelle layouts can vary dramatically, especially as technology advances and different approaches to power generation dictate changes.

Figure 25 Lattice tower.

5.2.4 Gearboxes

The gearbox, or transmission, of a turbine is connected to the rotor through the low-speed shaft. The power captured by the rotor is transferred to the generator through the high-speed shaft. In this case, high speed is a relative term

Figure 26 Tubular tower.

Figure 27 Tubular tower assembly.

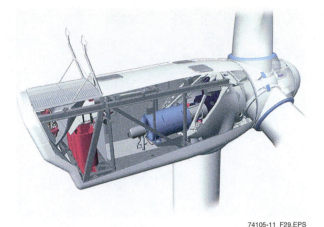

Figure 29 Vestas® V90 nacelle.

Figure 28 Turbine access through the tower.

since the high-speed shaft typically rotates below 2,000 rpm. The force and pressure placed on gearboxes is extreme due to the size and weight of the rotor. The external wind forces that are applied to the rotor are also placed on the gearbox. A typical gearbox is pictured in Figure 30.

Figure 30 Wind turbine gearbox.

Gearboxes are usually multi-stage, since the high-speed shaft of the gearbox must turn 25 to 50 times faster than the rotor. This application is a little unusual for a gearbox, as most gearboxes change high-speed inputs into lower speeds to achieve higher torque. Due to the stress imposed on gearboxes, they are commonly responsible for more maintenance and turbine downtime than all other major components.

On Site

Gearbox-Free Power Generation

Several companies have created turbine systems that eliminate the gearbox by using generators that can develop an equal amount of power at normal rotor speeds. The generator, however, must be much larger to accomplish the task. In spite of the larger size, the weight of the generator has been reduced using permanent magnets in the generator rotor in place of heavy copper-coil electromagnets. Not only is a high-maintenance component eliminated with this approach, but also the nacelle weight is reduced by 15 to 20 percent. While the initial reason was to eliminate the gearbox in offshore systems, continued success with this design will lead to their use in onshore units as well.

5.2.5 Electric Power Components

The major turbine power components include the generator, the pad-mounted transformer, and the converter.

The generator of a traditional HAWT unit is located in the nacelle and is turned by the gearbox. Generators that can provide power directly from the low-speed shaft of the rotor (without a gearbox) are becoming more popular. These systems are commonly referred to as direct-drive systems. Removing the gearbox eliminates the majority of major failures and high-cost repairs.

Older and smaller wind turbines may generate DC voltage to provide charging current for battery banks. However, it is far more practical for larger, grid-connected turbines to generate AC voltage. *Figure 31* shows an example of a typical commercial wind turbine generator. Turbine system generators typically produce 550 to 700 VAC power.

The grid operator determines the voltage for the grid interface. As increasing numbers of wind farms are started up and their output is tied to the electrical grid, electrical system operators must develop and enforce stringent standards for the power provided. In simple terms, utility grids need wind power supplies to behave much like any other power plant, although the two differ in very significant ways. In periods of light or nonexistent wind, the turbine generator cannot function. As turbine power drops off and later returns to the grid, special switching systems and components are required to avoid electrical circuit damage from the high voltage and current flow.

Power produced by the generator is sent down through the tower to a pad-mounted step-up transformer. There, the voltage is increased for transmission to a power substation. Since these transformers are quite heavy, they are mounted at ground level to minimize the weight of the nacelle.

The power leaving the AC generator varies in both voltage and frequency. AC converters (*Figure 32*) are used to clean up and condition the power at each turbine. The DOE awarded large research grants to the industry in 2010 for high-efficiency converters that would be combined with new transformer technology. The goal is to provide grid-ready power from a single component. This would eliminate the need for a pad-mounted transformer altogether.

5.2.6 HAWT Yaw and Pitch

As previously mentioned, HAWTs must be pointed into the wind for proper performance. This means that the turbine assembly must be able to turn on its vertical axis. This directional management is known as yaw control (*Figure 33*). Yaw control is only necessary for HAWTs; VAWTS are omnidirectional by design. Smaller turbines may self-control their yaw angle reasonably well using a simple tail vane. Larger units must be powered into position electrically or hydraulically. A small tail vane integrated into the turbine assembly still plays a role by providing sensory input to the yaw control mechanism. In most cases, the goal is to position the rotor perpendicular to the wind for maximum power production. However, there are occasions when the wind is simply too intense for the tur-

74105-11_F32.EPS

Figure 32 Power converter.

74105-11_F31.EPS

Figure 31 Wind turbine generator.

Multiple Generators

One manufacturer has taken a different approach to the problem of large generator systems and related service issues. The Clipper Liberty uses four smaller generators turned by four separate shafts from one gearbox. The smaller generators can be lowered to the ground individually for service or replacement by using tower-mounted hoists instead of large mobile or built-up cranes.

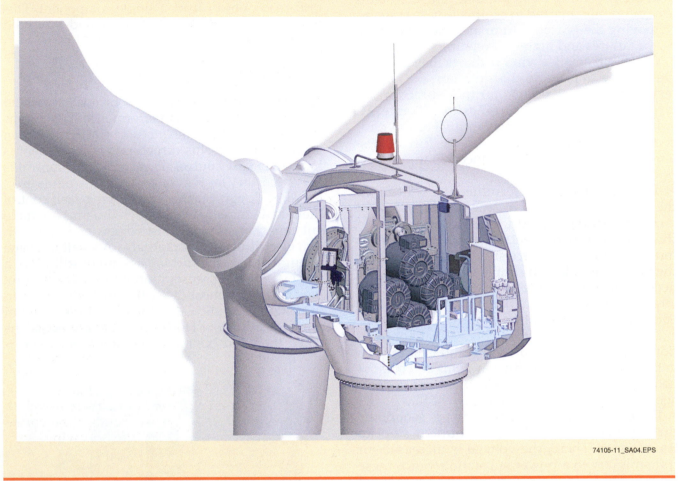

74105-11_SA04.EPS

bine system. The yaw control mechanism can assist by **furling** the rotor, or positioning the face away from a perpendicular wind strike. This slows the rotor speed and maintains the proper rpm for maximum power production without hurting the turbine. A fully furled rotor means the face of the rotor is parallel to the direction of the wind.

Pitch control provides another means of controlling the load applied to the rotor assembly. The pitch of an object is generally related to its rotation around a horizontal axis, such as pitching the nose of an aircraft up or down, just as yaw relates to the vertical axis. For wind turbines, it is related to the rotation of each blade along its **longitudinal** axis (*Figure 34*).

As in yaw control, electric motors or hydraulic actuators are generally used to change a blade's pitch. Pitch control works together with yaw control to improve performance of the turbine. Controlling the pitch also protects the blades and rotor from high winds by intentionally spoiling the aerodynamics and inhibiting lift.

5.2.7 Braking Systems

Wind turbines must be controlled in some way to prevent rotation at dangerous speeds. High winds could allow the rotor to rotate fast enough to simply self-destruct. In a matter of seconds, an uncontrolled rotor may exceed speeds that braking systems are able to counter before disaster occurs. It is essential that rotor speed be tightly controlled, and that backup systems be part of the design.

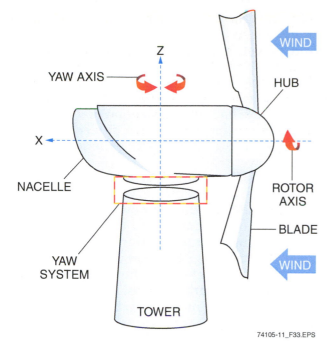

Figure 33 Yaw control.

Rotor braking is both aerodynamic and mechanical. When certain operating parameters indicate rotor speed is increasing too fast, the controller uses pitch and yaw adjustments to optimize performance and maintain the rotor at its best speed. Using mechanical braking systems for normal speed control would be very costly in

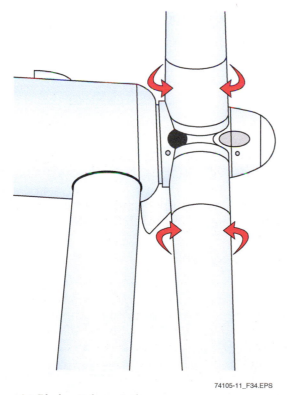

Figure 34 Blade pitch control.

terms of brake wear, especially when the wind does the braking for you.

When normal aerodynamic control fails and speed continues to rise, fail-safe systems begin to engage. Simple switches located in the rotor can open hydraulic valves to position blade **tip brakes** (*Figure 35*), spoiling the airflow across the blade. This represents another form of aerodynamic braking.

Mechanical braking is necessary for medium and large turbines as a back-up to aerodynamic braking. It also helps ensure the rotor cannot turn during service operations. Mechanical brakes are much like automotive disc brakes, with a rotor fixed to the shaft and a caliper-actuated brake pad. These brakes, however, are capable of operating at much higher temperatures than common automotive brakes. The huge mass of a turbine rotor is difficult to stop. Temperatures up to 700°C can be reached during a braking operation.

5.2.8 Supervisory Control and Data Acquisition (SCADA)

Wind turbines are often located in remote areas and operate for long periods without human attention. Virtually every function of a utility turbine must be monitored and controlled from a remote location. Electronic control systems must monitor operation to produce maximum power and protect the turbine from damage.

SCADA systems (*Figure 36*) are used to supervise, control, and collect data from a wind turbine or a collection of turbines in real time. A vast array of information is collected through sensor inputs from the turbine and the site. Some of the inputs to the SCADA system include the following:

- Wind velocity and direction
- Blade pitch
- Rotor rpm and yaw direction
- Oil and individual bearing temperatures
- Vibration levels
- Generator output

As the system uses this data to control turbine functions, vital information is collected and saved to provide trend reports. These reports assist in predictive maintenance and future programming schemes. For example, small increases in the temperature of a bearing over a period of time comprise a trend toward possible failure. The SCADA system can alert an operator thousands of miles away to the need for attention.

Figure 35 Tip brakes deployed.

6.0.0 SMALL WIND

The AWEA defines the small wind market as turbines which produce less than or equal to 100 kW of power. They are generally used for residential or light commercial applications. HAWTs in this class typically have a blade length of 10 m (32+ feet) or less.

6.1.0 Off-Grid Systems

Smaller wind turbines, much like solar photovoltaic systems, can be used to provide power to small facilities. Such systems can be valuable in areas beyond the reach of the grid. Technically,

grid power can reach the most remote location, but it is not economically feasible in many cases. Wind power, sometimes combined with solar power, can provide for basic electrical needs.

The problem, of course, is that wind power is not available at all times. Production is very good on some days and non-existent on others. To provide a more consistent source of electricity, the turbine is often used to charge batteries with any unused power.

Figure 37 shows a simple wind-powered system with batteries and an optional generator. When the wind turbine is producing power, it can be used in one of two ways. When an electrical demand exists, the power is routed directly to the **inverter** (*Figure 38*), where its DC power output is changed to usable AC power. If all the electrical power produced by the turbine is not consumed in the house or building immediately, the extra power is directed through a **charge controller** (*Figure 39*) to the battery bank. The batteries provide a means of storing power for later use. The charge controller ensures that the batteries are charged in a manner that maximizes their capacity and longevity. When power is needed and the wind turbine is unable to function, power is drawn from the batteries through the inverter instead. The optional generator is typically powered by a fossil fuel such as gasoline, diesel fuel, or propane. It can be used when the demand for power is exceptional or when wind power production is poor and the batteries have lost their charge.

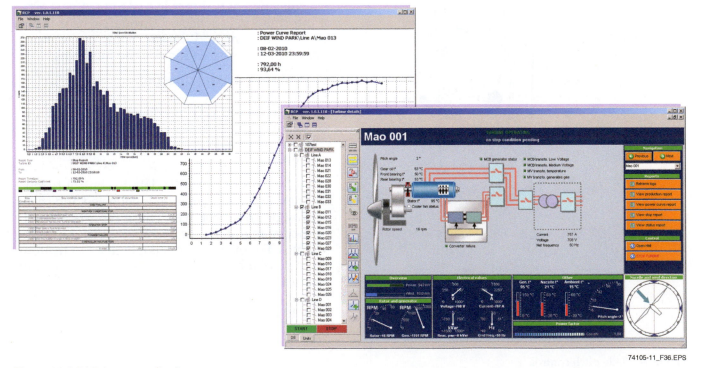

Figure 36 SCADA system displays.

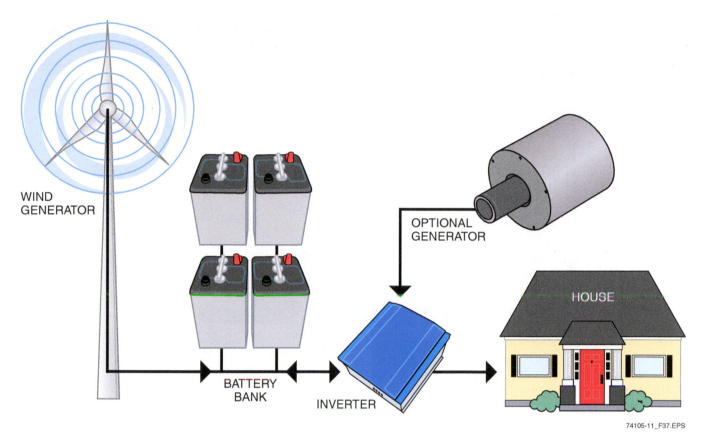

WIND GENERATOR

OPTIONAL GENERATOR

HOUSE

BATTERY BANK

INVERTER

74105-11_F37.EPS

Figure 37 Typical off-grid wind-powered system.

6.2.0 Grid-Tied Systems

Small wind grid-tied systems (*Figure 40*) are wind turbines that are installed along with commercial power. As a result, the wind turbine size is quite flexible. These systems can provide all the power needed for a structure, or only a fraction of it. They can also provide more power than needed, with the excess power sent upstream to the utility grid.

With **net metering** available in many locations, the electric meter actually deducts energy consumption from the register when the turbine is generating extra power and supplying it to the grid. When power is being used from the grid, the meter operates normally. At the end of each month, the net result could be in favor of either

74105-11_F38.EPS

Figure 38 Power inverter.

74105-11_F39.EPS

Figure 39 Charge controller.

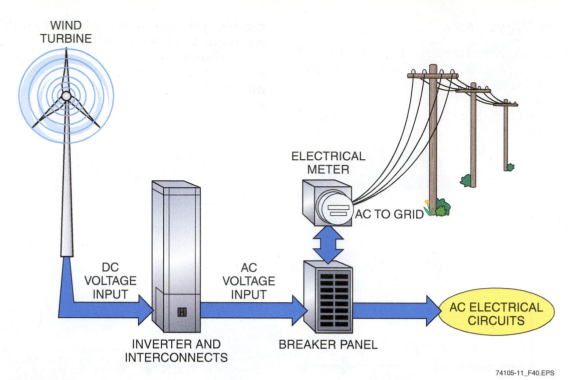

WIND
TURBINE

ELECTRICAL
METER

AC TO GRID

DC
VOLTAGE
INPUT

AC
VOLTAGE
INPUT

AC ELECTRICAL
CIRCUITS

INVERTER AND
INTERCONNECTS

BREAKER PANEL

74105-11_F40.EPS

Figure 40 Grid-tied wind turbine system.

party. Wind system owners may receive money or credit from the utility, if the turbine has generated more power than was used.

In spite of this obvious advantage, each approach to grid-tied systems has some disadvantages that must be carefully considered. When grid-tied systems are installed without a battery bank and utility power is interrupted for any reason, the wind turbine must be automatically disconnected. This must be done for safety purposes, to prevent power line workers servicing the grid from being shocked by a circuit they think is without power. The result, of course, is a complete loss of power to the structure. When battery banks are employed, the array is disconnected from the grid but the battery system remains connected to the

load through inverters designed specifically for this purpose. Battery power stays isolated from the grid.

Grid-tied systems without batteries are simple in design, have fewer components, and are less expensive to install. Not only is the initial cost with batteries much higher, but also the life cycle cost grows due to the added maintenance of the additional equipment. However, without batteries, there is no power when grid power is lost.

In some applications, the need for an uninterruptable power supply cannot be avoided. However, when that is not a necessity, the simplicity and reduced cost of the grid-tied system without batteries is very attractive financially.

Big Horn Wind Farm

GOING GREEN

The Big Horn wind farm is populated with 133 GE turbines, each rising 119 meters (389 feet) from the ground to the top of the blade tips. Yet the population below has changed very little, in spite of the new 200-ton residents. Of the 15,000 acres dedicated to the project, only a small percentage of the ground area is occupied. The rest of the area remains in use for wheat farming and ranching, just as before. The project also invested in the environment by conserving 455 acres south of the farm.

In the nearby town of Bickleton, known as the bluebird capital of the world, funding from the project replaced 250 bluebird boxes. Research did not indicate that the turbines present a danger to the bluebirds; this was done to improve the area. Iberdrola Renewables, the owner and operator, continues to support the local environment by developing community education programs about native birds.

7.0.0 THE WIND FARM

In order to coordinate the production of wind power on a utility-scale and its use on the power grid, wind turbines are gathered together in wind farms. Areas dedicated to groups of turbines can also be referred to as wind parks, wind industrial parks, or wind energy sites. One good example is the Big Horn wind farm (*Figure 41*), built in the state of Washington in 2007. This particular farm generates roughly 200 MW using 133 GE 1.5 MW turbine systems. As you can see, the turbines here are aligned in lines or strings.

As shown in *Figure 42*, power generated by each turbine is routed down the tower to a pad-mounted transformer. The substation collects power from the turbine strings and then boosts the voltage even higher for long distance transmission. The substation must match the voltage on the grid. Grid voltages can vary quite a bit, from 110 kV to as much as 765 kV.

74105-11_F41.EPS

Figure 41 Big Horn wind farm.

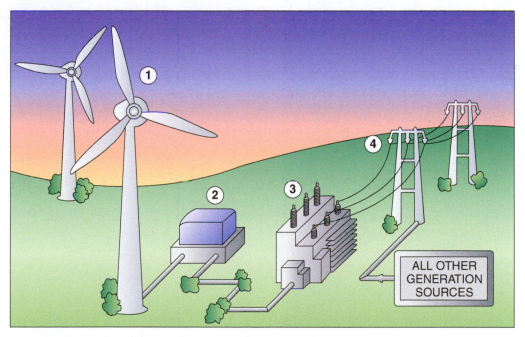

① Rotating generator converts wind energy to electricity.
② Transformer increases voltage for transmission to substation.
③ Substation increases voltage for transmission over long distance.
④ Transmission to the grid.

74105-11_F42.EPS

Figure 42 Turbine power delivery to the utility grid.

The most desirable wind farm sites are where wind resources are the greatest. An average annual wind speed of at least 11 mph is generally needed. Other aspects to be considered prior to development include the following:

- The maximum power-generating capacity that can be installed based on site size and grid connection issues
- Required setbacks from existing structures or protected areas
- The location of important sight lines and their relationship to the turbines
- The location of dwellings that could be impacted by blade flicker or shadows
- An analysis of the power buyer and the associated market

Each turbine location must be well planned before construction. Turbines themselves represent obstructions to the wind and can create turbulence that may negatively affect other turbines. Manufacturers must make guarantees regarding performance and output, so the wind farm designer must adhere to their specifications of placement. Special software packages are available to assist in layout with computerized modeling of the facility.

At the end of 2010, the world's largest wind site was the Roscoe Wind Farm (*Figure 43*) in Roscoe, TX. With a total capacity of 781.5 MW, it surpassed its nearby neighbor, the Horse Hollow Wind Energy Center. It has a capacity of about 735 MW. Cotton and wheat farming continues at the feet of the turbine army. Farmers also collect substantial lease payments from the site operators. There is no doubt that its capacity will be surpassed by another project as the use of wind power continues to grow and expand.

74105-11_F43.EPS

Figure 43 Roscoe Wind Farm.

SUMMARY

The wind industry is already a key component in the expansion of alternative energy in the United States and abroad. The long history of wind power has provided guidance to an industry that hopes to achieve 20 percent of US power from wind energy by 2030.

The science of wind and its energy is well tested and understood, but consistency and perfect performance will likely always remain elusive. Performance has proven, for now, that the three-bladed, horizontal axis rotor is best for medium- and large-scale wind turbines. The performance of a HAWT is further optimized by controlling the yaw of the rotor assembly and the pitch of each blade.

Small wind systems continue to advance as well. Such systems can be used in off-grid applications for reduced energy independence. Combining small wind power with the grid offers the stability of a reliable power source with a renewable source, potentially resulting in significant savings for the user. As wind energy sites continue to be developed, both on- and offshore, reaching the 2030 goal becomes a real possibility.

1. One of the distinct advantages to the use of wind power is that _____.

   a. wind turbines are very inexpensive to purchase and install
   b. wind energy can reduce our dependence on fossil fuels and those who control fossil-fuel resources
   c. wind energy can easily replace all fossil fuels as a source of electricity by 2030
   d. a sufficient amount of wind for power production is available everywhere

2. As of 2010, what state led the United States in total installed wind energy capacity?

   a. Texas
   b. Iowa
   c. California
   d. North Dakota

3. Which organization has the mission of promoting wind power growth through advocacy, communication, and education?

   a. DOE
   b. NREL
   c. AWEA
   d. NOAA

4. Potential energy is a measure of the energy within an object or mass that results from being set in motion.

   a. True
   b. False

5. The two primary factors that determine the energy carried by the wind are its mass and its _____.

   a. velocity
   b. density
   c. pressure
   d. temperature

6. Doubling the diameter of a horizontal axis rotor would have what effect on its swept area?

   a. It doubles the area.
   b. It triples the area.
   c. It quadruples the area.
   d. It quintuples the area.

7. The theoretical limit of how much wind energy a rotor could capture is known as the _____.

   a. Betz limit
   b. kinetic limit
   c. Hutter limit
   d. potential limit

8. Due to the enormous size of the largest turbines, wind sites for them are developed _____.

   a. offshore
   b. in other countries
   c. near areas of high electrical load
   d. in rugged, remote mountain passes

9. What axis of control does yaw refer to?

   a. Lateral
   b. Vertical
   c. Horizontal
   d. Longitudinal

10. Mechanical braking is the normal means of controlling rotor speed.

    a. True
    b. False

Name: _____

Date: _____

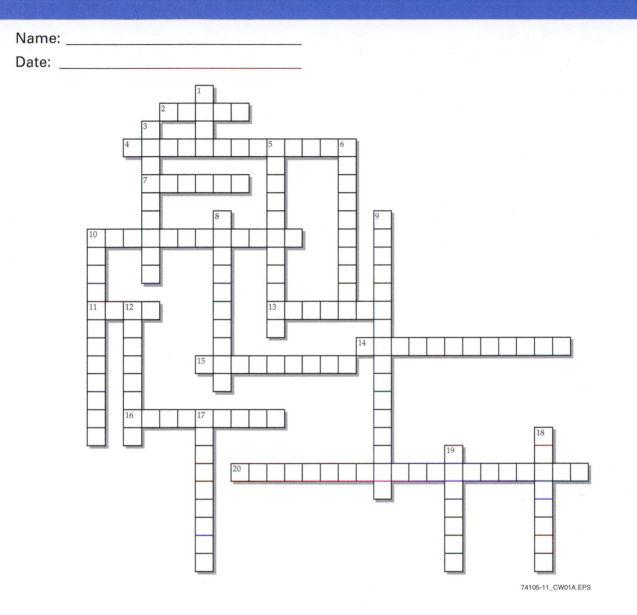

74105-11_CW01A.EPS

Across:

2. Computerized system to monitor and control turbines and wind energy sites

4. Energy stored within a body from its motion

7. An apparatus that converts mechanical energy into electrical energy

10. A means of quantifying the available wind energy

11. Wind turbine system with a horizontal axis

13. Streamlined housing or enclosure for a turbine system's main components

14. Running lengthwise, or extending along the length of an object

15. Equals 59.3 percent of the mathematically available wind power

16. Turbine rotor blades pass through this

20. The surface, often elevated, that the wind is passing smoothly across

Down:

1. Wind turbine system with a vertical axis

3. Describes the variations of wind speed at different heights above Earth and obstacles

5. Approach used to track power flow both in and out of a consumer location

6. Managing the direction a wind turbine rotor faces

8. A device that measures wind velocity

9. A device that regulates DC power from an energy source to regenerate batteries

10. Adjusting the longitudinal position of individual turbine blades

12. A circular graph documenting the history of wind speed and frequency for a given location

17. Rotor blade ends that can rotate independently and spoil airflow

18. A device that changes DC power to AC power

19. Prevents excessive rotor speed by turning the rotor away from a direct wind facing

Tanzania Adams

Project Manager, Power Generation Resource
Policy & Planning Georgia Power, Atlanta GA

How did you get started in the power or energy industry?
I began in the power industry right out of college as a reliability/planning engineer, then as a field engineer in power delivery with Alabama Power Company. However, my interest in renewables began in college when I was president of the student chapter of the National Society of Black Engineers (NSBE). We hosted an environmental summit/conference at the University of Alabama. Renewable energy was an important part of the discussions.

Who inspired you to enter the industry?
No one person inspired me to enter the industry. Power and energy are at the cornerstone of all industry. They are absolutely essential to maintaining the quality of life that we, as Americans, enjoy.

What do you enjoy most about your job?
I enjoy the variety inherent in working on projects dealing with leading-edge technologies. In planning, we are often looking 5 to 10 years into the future to make sure we have the resources in place to meet our forecasted needs. In addition, in other roles within the power industry, I have had the good fortune to work with many of our customers who were trying to integrate the latest and most efficient technologies into their business processes. By working in this industry, I had the privilege to be on their team to bring those improvements to life.

Do you think training and education are important in the power and energy industry? If so, why?
Training and education are crucial in the power and energy industry. Energy must be respected and handled with extreme caution and care. Thus, the proper training and education are needed to ensure a safe outcome.

How important are NCCER credentials to your career?
NCCER credentials can be utilized with the individuals on the project teams that I work with. The type of training that NCCER provides can assist everyone from the lineworkers, who construct the power delivery designs I create, to the electricians working with the large industrial/commercial customers we serve.

Would you suggest power and energy as a career to others? If so, why?
I would definitely suggest this industry to others. Energy is at the cornerstone of all industry and business. Moreover, energy is essential to maintaining our way of life. Therefore, there will always be a need for someone with knowledge, skill, and experience in the energy industry.

How do you define craftsmanship?
I define craftsmanship as the skillful ability to implement solutions.

Trade Terms Introduced in This Module

Anemometer: A device used to measure wind velocity that often incorporates wind direction as well.

Betz limit: The theoretical limit of 59.3 percent of the available wind power that a rotor can capture. The theory is named for its German developer, Albert Betz.

Charge controller: A device that controls the DC power from an energy source used to charge batteries, to ensure the batteries charge to their maximum levels.

Dynamo: An apparatus that converts mechanical energy into electrical energy, typically in the form of direct current.

Effective ground level: The actual surface that air movement is passing across, as opposed to actual ground level. For example, the effective ground level for a wind blowing over a dense forest would be the tops of the trees rather than the ground itself.

Furling: One method of preventing excessive wind turbine rotor speed through yaw control by turning the rotor blades away from a direct wind facing.

Horizontal-axis wind turbines (HAWTs): A wind turbine that spins on an axis which is horizontal or nearly so, much like the early windmills of the western and mid-western United States Also referred to as a conventional turbine or propeller-style, they are directional by design; that is, the rotor must face into the wind for maximum performance.

Inverter: An electronic device that inverts a DC power source to an AC power supply.

Kinetic energy: The energy contained in a mass or body caused by its motion.

Longitudinal: Running lengthwise, or extending along the length of an object. A line drawn the length of an object would indicate its longitudinal axis.

Nacelle: A streamlined housing or enclosure that contains the major working components of the turbine system at the top of the tower.

Net metering: The metering approach used to accommodate renewable energy sources that are on the load or customer side of the grid, allowing power to flow in both directions and the costs credited accordingly. The meter typically records flow in both directions.

Pitch control: Controlling the longitudinal position of turbine blades as a means of rotor speed control and braking.

Power density: The means of quantifying the power available in wind per unit of area, generally expressed as watts per square meter (w/m^2); in English units, it is expressed as watts per square foot (w/ft^2).

Supervisory control and data acquisition (SCADA) system: A computerized system used to supervise, control, monitor, and collect historical data from an individual wind turbine or a collection of turbine systems using real-time information and commands.

Swept area: The area that turbine rotor blades pass through.

Tip brakes: Rotor blade tips designed to rotate independent of the rest of the blade, allowing it to spoil the aerodynamic characteristics and reduce rotor speed.

Vertical-axis wind turbines (VAWTs): A wind turbine with a rotor that spins on a vertical or near-vertical axis. VAWTs are generally omnidirectional and allow for the drive train and generators systems to be mounted at ground level.

Wind rose: A circular graph that depicts the frequency at which winds blow from a given direction at a given location, generally reported as a percentage of time. Wind roses may also contain other information through the use of color, such as how often the wind blows from a direction at a given velocity.

Wind shear: The wind velocity variations that occur at different heights above the earth.

Yaw control: Management of a wind turbine's facing direction by rotation of the turbine assembly on its vertical axis.

Additional Resources

This module presents thorough resources for task training. The following resource material is suggested for further study.

American Wind Energy Association (AWEA). www.awea.org.
Introduction to Wind Principles. Thomas E. Kissell. Boston, MA: Prentice Hall.
Wind Power. Paul Gipe. White River Junction, VT: Chelsea Green Publishing Company.
US Department of Energy, Wind Powering America Program. www.windpoweringamerica.gov.

Figure Credits

Courtesy of DOE/NREL, Module opener, Figures 2, 6, 12, 16, 24, 30, and 41

Brian Shultz, Figures 1, 26-28, and 35

©iStockphoto.com/RelaxFoto.de, Figure 3

Portrait of Charles F. Brush, Sr. ca. 1900.Box 40, Folder 4: The Charles F. Brush, Sr., Papers. Copy of an original in the Special Collections Research Center, Kelvin Smith Library, Case Western Reserve University, Cleveland, Ohio., Figure 4 (top photo)

Brush windmill ca.1890-1900. Box 41, Folder 8: The Charles F. Brush, Sr., Papers. Copy of an original in the Special Collections Research Center, Kelvin Smith Library, Case Western Reserve University, Cleveland, Ohio., Figure 4 (bottom photo)

Courtesy of Steve Rossi, Figure 5

Photographer: Eric Jacobson, Permission by: Michael D. Zuteck, Figure 8

National Weather Service Eastern Region Headquarters, SA01 and SA02

APRS World, LLC, SA03

National Weather Service Weather Forecast Office, Chicago, IL, Figure 13

WePOWER LLC, Figure 14

©iStockphoto.com/chiyacat, Figure 15

WindBlue Power, Figure 18

US Department of Energy, Office of Energy Efficiency and Renewable Energy, Figure 19

©iStockphoto.com/thad, Figure 20

©Ralf Nöhmer – Fotolia.com, Figure 21

Lonzelle Noack, Figure 22

Photo courtesy of Paul Cryan, US Geological Survey, Figure 23

©2011 Photos.com, a division of Getty Images. All rights reserved., Figure 25

Photo courtesy Vestas Wind Systems, Figure 29

ABB Inc., Figures 31 and 32

Courtesy of Clipper Windpower, Inc., SA04

Courtesy of DEIF Wind Power Technology, Figure 36

SMA Solar Technology AG, Figure 38

Courtesy of Green Mountain Energy Company, Figure 42

E.ON Climate & Renewables, Figure 43

Southeast Regional Climate Center, Project 1

Courtesy of GreenLearning Canada Foundation, www.GreenLearning.ca, Project 2

KidWind Project www.kidwind.org, Project 3

NCCER CURRICULA — USER UPDATE

NCCER makes every effort to keep its textbooks up-to-date and free of technical errors. We appreciate your help in this process. If you find an error, a typographical mistake, or an inaccuracy in NCCER's curricula, please fill out this form (or a photocopy), or complete the online form at **www.nccer.org/olf**. Be sure to include the exact module ID number, page number, a detailed description, and your recommended correction. Your input will be brought to the attention of the Authoring Team. Thank you for your assistance.

Instructors – If you have an idea for improving this textbook, or have found that additional materials were necessary to teach this module effectively, please let us know so that we may present your suggestions to the Authoring Team.

NCCER Product Development and Revision

13614 Progress Blvd., Alachua, FL 32615

Email: curriculum@nccer.org
Online: www.nccer.org/olf

❏ Trainee Guide   ❏ AIG   ❏ Exam   ❏ PowerPoints   Other _____

Craft / Level: _____   Copyright Date: _____

Module ID Number / Title: _____

Section Number(s): _____

Description: _____

_____

_____

_____

Recommended Correction: _____

_____

_____

_____

_____

Your Name: _____

Address: _____

_____

Email: _____   Phone: _____

Glossary

Air mass: The thickness of the atmosphere that solar radiation must pass through to reach Earth.

Alternative energy: Energy that is provided by means not related to fossil fuels. Most forms of alternative energy are considered renewable, but not all.

Altitude: The angle at which the sun is hitting the array.

Ambient temperature: The air temperature of an environment.

Amorphous: A low-efficiency type of photovoltaic cell characterized by its ability to be used in flexible forms. Also known as thin film.

Anaerobic digester: Equipment used to recover methane and other byproducts by using bacteria to break down organic matter in an oxygen-free envireoment.

Anemometer: A device used to measure wind velocity that often incorporates wind direction as well.

Array: A complete PV power generating system including panels, inverter, batteries and charge controller (if used), support system, and wiring.

Autonomy: The number of days a fully charged battery system can supply power to loads without recharging.

Azimuth: For a fixed PV array, the azimuth angle is the angle clockwise from true north that the PV array faces.

Backfeed: When current flows into the grid.

Balance of system (BOS): The panel support system, wiring, disconnects, and grounding system that are installed to support a PV array.

Base load unit: An electrical power-generating unit that has the primary mission of supporting basic power needs and typically operates on a full-time basis. A base load is considered the lowest amount of electrical power that is required to satisfy consumer needs at any time.

Betz limit: The theoretical limit of 59.3 percent of the available wind power that a rotor can capture. The theory is named for its German developer, Albert Betz.

Biofuel: A fuel that originates from an organic, renewable source, such as ethanol or methane.

Biogenic: Produced by living organisms, such as methane.

Biomass: Fuel that originates from living or recently living organic matter, generally with little or no processing before use.

Boiling water reactor (BWR): A light water nuclear reactor in which the water boils in the reactor core and is drawn up through a separator in the top of the pressure vessel. The water is separated from the vapor, which is sent directly to a steam turbine.

Breeder reactor: A reactor that creates more fissionable material than it uses.

Brownout: A temporary decrease in grid output voltage typically caused by peak load demands.

Building-integrated photovoltaics (BIPV): A PV system built into the structure as a replacement for a building component such as roofing.

Cellulose: The fibrous carbohydrate found in the walls of green plants cells; cellulose gives strength and rigidity to plants.

Centralized power generation: The generation of electric power in large volumes at a single location to serve many consumers. Most locations fitting this description are fossil fuel, nuclear, or hydroelectric facilities.

Charge controller: A device that controls the DC power from an energy source used to charge batteries, to ensure the batteries charge to their maximum levels.

Charge controller: A device used to regulate the charging and discharging of the battery system to prevent overcharge and excess discharge.

Cogeneration: The simultaneous generation of usable electric power and heat from a single source or process; also known as combined heat and power (CHP).

Combined heat and power (CHP): The simultaneous generation of usable electric power and heat from a single source or process; also known as cogeneration.

Combiner box: A junction box used to connect strings of solar panels to create a larger array, and to provide a convenient array disconnect point.

Concentrated solar thermal (CST): The process of generating energy by concentrating the sun's power to a single point, heating liquids to create steam or high-temperature liquids. The medium can be stored to allow for power generation when solar power is not available.

Concentrating collector: A device that maximizes the collection of solar energy by using mirrors or lenses to focus sunlight onto a central receiver.

Convection: The transfer of heat through a gas or liquid by distribution of heat through currents of air circulating within a closed area.

Core: The central part of a nuclear reactor where fission occurs.

Cropland: The total area of land used for crops, pasture, and idle land. Idle land is land that has been cultivated but is now unused.

Declination: The angle between the equator and the rays of the sun.

Depth of discharge (DOD): A measure of the amount of charge removed from a battery system.

Distributor plate: The supporting plate or grate in the bottom of a gasifier, perforated by nozzles, holes, or other pathways that allow air penetration into the feedstock resting on the plate.

Doped: A material to which specific impurities have been added to produce a positive or negative charge.

Dual-axis tracking: An array mounting system designed to adjust both the horizontal and vertical axes of a panel to precisely follow the movement of the sun.

Dynamo: An apparatus that converts mechanical energy into electrical energy, typically in the form of direct current.

Effective ground level: The actual surface that air movement is passing across, as opposed to actual ground level. For example, the effective ground level for a wind blowing over a dense forest would be the tops of the trees rather than the ground itself.

Electric grid: The electrical infrastructure that transmits and distributes power across the United States.

Electrochemical solar cell: A type of PV cell that replaces silicon with a light-sensitive dye that absorbs light and produces current.

Elevation: A measure of a location's relative height in reference to sea level.

Evacuated tube collector: A type of solar thermal heat collector consisting of parallel rows of transparent glass tubes with a vacuum between the tubes to preserve heat.

Fission: The splitting of atoms, which releases a tremendous amount of energy in the form of heat.

Flat plate collector: A type of solar thermal heat collector consisting of metal box with a transparent cover that contains three main components: an absorber to collect the heat, tubes containing water or air, and insulation.

Fracturing: A mining process used to break or separate rock, usually by injecting fluid into a hole, either natural or manmade, under high pressure.

Fuel cell: A device that harnesses the energy produced by a chemical reaction between hydrogen and oxygen to produce direct current.

Fuel cells: Devices that generate electrical power by using the chemical energy released by a fuel and oxidant reaction.

Furling: One method of preventing excessive wind turbine rotor speed through yaw control by turning the rotor blades away from a direct wind facing.

Fusion: The process in which atomic nuclei collide so fast they stick together and emit a large amount of energy. In the center of most stars, hydrogen fuses into helium.

Gasification: Any chemical or heat process that results in changing a substance into a gas. Gasification is used to transform biomass and other solids into a useful gas, commonly known as syngas.

Greenhouse gases: Gases that contribute to the Earth's warming by trapping and reflecting solar radiation and heat back towards the planet. CO_2 and ozone are the two primary greenhouse gases.

Grid-connected system: A PV system that operates in parallel with the utility grid and provides supplemental power to the building or residence. Since they are tied to the utility, they only operate when grid power is available. Also known as a grid-tied system.

Grid-interactive system: A PV system that supplies supplemental power and can also function independently through the use of a battery bank that can supply power during outages and after sundown.

Grid-tied system: See *grid-connected system*.

Gross Domestic Product (GDP): The final value of all goods and services produced within the borders of a country during some period. Several methods can be used to determine the total value. Government, or public spending, is often separated from all other spending.

Heavy water: Water laced with a large number of molecules that contain deuterium atoms; deuterium is also called heavy hydrogen.

Heavy water reactor: In most reactors, the moderator is plain water; but some reactors use heavy water, which does not absorb neutrons as regular hydrogen does, making it useful in slowing neutrons from fission.

Heliostats: A group of sun-tracking mirrors used to focus the sunlight on a central receiver in a solar thermal power plant.

Horizontal-axis wind turbines (HAWTs): A wind turbine that spins on an axis which is horizontal or nearly so, much like the early windmills of the western and mid-western United States. Also referred to as a conventional turbine or propeller-style, they are directional by design; that is, the rotor must face into the wind for maximum performance.

Hybrid system: A grid-interactive system used with other energy sources, such as wind turbines or generators.

Insolation: The equivalent number of hours per day when solar irradiance averages 1,000 W/m^2. Also known as peak sun hours.

Integral-collector storage system: A type of solar thermal system consisting of one or more storage tanks in an insulated box with a glazed side facing the sun.

Interconnection agreement: A contract between a utility and a power provider that outlines the specific means of connection and how both parties will be compensated.

Inverter: A device used to convert direct current to alternating current.

Inverter: An electronic device that inverts a DC power source to an AC power supply.

Irradiance: A measure of radiation density at a specific location.

Kilowatt (kW): One thousand watts.

Kilowatt-hour (kWh): The primary unit used to bill for the consumption of electric power. For example, one kWh is equal to the use of one kW of power for one hour, or the use of two kW of power for 30 minutes.

Kinetic energy: The energy contained in a mass or body caused by its motion.

Latitude: A method of determining a location on Earth in reference to the equator.

Light water reactor: A nuclear reactor that uses plain water as a coolant and moderator.

Longitudinal: Running lengthwise, or extending along the length of an object. A line drawn the length of an object would indicate its longitudinal axis.

Maximum power point tracking (MPPT): A battery charge controller that provides precise charge/discharge control over a wide range of temperatures.

Megawatt (MW): One million watts.

Metabolic: Refers to processes that make energy and matter available to cells. Any reaction in a living organism that builds and breaks down organic molecules and produces or consumes energy in the process is metabolic.

Microbes: Very tiny living organisms, visible through a microscope, such as a bacteria, funguses, or viruses.

Microgrid: A small power grid that typically includes several forms of power generation and the primary users.

Module: A PV system component consisting of numerous electrically and mechanically connected PV cells encased in a protective glass or laminate frame. Also known as a PV panel.

Monocrystalline: A type of PV cell formed using thin slices of a single crystal and characterized by its high efficiency.

Municipal solid waste (MSW): The kind of garbage found in a landfill.

Nacelle: A streamlined housing or enclosure that contains the major working components of the turbine system at the top of the tower.

Net metering: A method of measuring power used from the grid against PV power put into the grid.

Net metering: A system of monitoring power coming into the grid as well as grid power used by a consumer. In most cases, the electric meter can turn in either direction.

Net metering: The metering approach used to accommodate renewable energy sources that are on the load or customer side of the grid, allowing power to flow in both directions and the costs credited accordingly. The meter typically records flow in both directions.

Off-grid system: A PV system typically used to provide power in remote areas. Off-grid systems use batteries for energy storage as well as battery-based inverter systems. Also known as a standalone system.

Parabolic cooling tower: An upright cooling tower with a "waist." The shape induces a natural draft, drawing air in at the bottom and forcing it out at the top.

Peak sun hours: See *insolation*.

Peaking load unit: An electrical power-generating unit used to satisfy electrical demands that are above the volume considered the base load. Peaking load units are operated intermittently as needed.

Phantom loads: Devices that use electrical power even when they are not turned on or actively engaged in their design function.

Photons: An individual particle of electromagnetic energy such as light; a basic unit that has no mass.

Photovoltaic (PV): Describes the process of transforming light energy into electrical power.

Pitch control: Controlling the longitudinal position of turbine blades as a means of rotor speed control and braking.

Polycrystalline: A type of PV cell formed by pouring liquid silicon into blocks and then slicing it into wafers. This creates non-uniform crystals with a flaked appearance that have a lower efficiency than monocrystalline cells.

Power density: The means of quantifying the power available in wind per unit of area, generally expressed as watts per square meter (w/m^2); in English units, it is expressed as watts per square foot (w/ft^2).

Pressurized water reactor (PWR): A type of light water reactor is which the water is kept under pressure and not allowed to expand, so it gets hotter and hotter but stays in its liquid state, creating enormous pressure. The superheated, pressurized water is carried to a steam generator where it is used to produce steam.

Pulse width-modulated (PWM): A control that uses a rapid switching method to simulate a waveform and provide smooth power.

Pyrolysis: A process in which biomass is subjected to heat greater than 400°F to remove the volatile matter and change the remaining material to charcoal and oil. Volatile matter is the material given off as vapor or gas, but it does not include water vapor.

Renewable energy: Energy that comes from a source that is naturally renewed or sustained. Renewable energy sources are also considered alternative energy sources.

Renewable Portfolio Standard (RPS): Also known as a Renewable Electricity Standard at the federal level. An RPS outlines the state-required production or procurement of power generated from renewable energy sources by utility companies and other power providers.

Roentgens per man (rems): A quantity that relates to the amount of damage to the human body an absorbed dose of radiation could (but does not necessarily) cause. It is often expressed in thousandths of a rem, or millirems.

Sea level: A measure of the average height of the ocean's surface between low and high tide. Sea level is used as a reference for all other elevations on Earth.

Semiconductor: A material that exhibits the properties of both a conductor and an insulator.

Sine wave: A form presented as a ripple, representing the consistent frequency and amplitude of electrical power.

Single-axis tracking: An array mounting system designed to adjust either the horizontal or the vertical axis of a panel to follow the movement of the sun.

Smart grid: A digitally enhanced grid system to allow greater control of both loads and power resources.

Solar photovoltaic (PV) system: A power production system that converts sunlight into electricity using a semiconductor.

Solar thermal energy (STE): Energy that uses solar power to generate heat.

Solar thermal system: A system that uses sunlight to heat air or water.

Spectral distribution: The distortion of light through Earth's atmosphere.

Standalone system: See *off-grid system*.

Standard Test Conditions (STC): Standardized panel ratings based on a specific operating temperature, solar irradiance, and air mass.

Stover: The leftover parts of corn after harvest, including stalks, leaves, and husks.

Sun path: The sun's altitude and azimuth at various times of year for a specific location or latitude band.

Superconductors: Elements or metallic alloys that are capable of conducting electrical power with little or no resistance to flow, eliminating losses. Typically, these materials function this way at a temperature near absolute zero; however, new discoveries of materials that exhibit this characteristic well above absolute zero give hope for future materials that can do so at normal temperatures.

Supervisory control and data acquisition (SCADA) system: A computerized system used to supervise, control, monitor, and collect historical data from an individual wind turbine or a collection of turbine systems using real-time information and commands.

Swept area: The area that turbine rotor blades pass through.

Tertiary: Belonging to the third level or order.

Thermosiphon system: A type of solar thermal system consisting of a tank mounted above one or more flat panel collectors. Thermosiphon systems rely on convection to circulate water through the collectors and to the tank.

Thin film: See *amorphous*.

Tilt angle: The position of a panel or array in reference to horizontal. Often set to match local latitude or in higher-efficiency systems, the tilt angle may be adjusted by season or throughout the day.

Tip brakes: Rotor blade tips designed to rotate independent of the rest of the blade, allowing it to spoil the aerodynamic characteristics and reduce rotor speed.

Utility-scale: Power generation on a scale that is usable and significant to utility companies. The power may be generated by the utility's own systems or by others who sell power to the utility directly. Utility-scale power generating systems are typically listed with the DOE as such a facility.

Utility-scale solar generating system: Large solar farms designed to produce power in quantities large enough to operate a small city.

Vertical-axis wind turbines (VAWTs): A wind turbine with a rotor that spins on a vertical or near-vertical axis. VAWTs are generally omnidirectional and allow for the drive train and generators systems to be mounted at ground level.

Vitrification: The conversion of a material into extremely hard glass by subjecting it to temperatures lower than its normal melting point.

Watt: An SI unit of power measurement equal to the power produced when a current of one ampere flows at a potential electrical difference of one volt. One watt is equal to one joule per second.

Watt-hours (Wh): A unit of energy typically used for metering.

Wind rose: A circular graph that depicts the frequency at which winds blow from a given direction at a given location, generally reported as a percentage of time. Wind roses may also contain other information through the use of color, such as how often the wind blows from a direction at a given velocity.

Wind shear: The wind velocity variations that occur at different heights above the earth.

Yaw control: Management of a wind turbine's facing direction by rotation of the turbine assembly on its vertical axis.

Zero energy district: An area that seeks to both reduce power consumption and generate all needed power from local alternative sources.

Index

V

VAWTs (Vertical-axis wind turbines). *See* Vertical-axis wind turbines (VAWTs)
Vazquez, Tony, (Module 3): 39–40
Velocity (V), wind
    calculating, (Module 5): 8–9
    factors affecting, (Module 5): 8, 10
    height and, (Module 5): 9–10
Venus of Dolní Vestonice, (Module 2): 18
Vermont, (Module 5): 4
Vertical-axis wind turbines (VAWTs), (Module 5): 15, 36
Very high temperature reactor (VHTR), (Module 3): 33
Vestas® V90 nacelle, (Module 5): 22
VHTR (Very high temperature reactor). *See* Very high temperature reactor (VHTR)
Vitrification, (Module 3): 22, 41
Volatiles, (Module 2): 11–12
Volt (V), (Module 4): 12
Volt-amp (VA), (Module 4): 12

W

WACs (Wind Application Centers). *See* Wind Application Centers (WACs)
Walton, Ernest, (Module 3): 27
Waste generation. *See* Landfills; Municipal solid waste (MSW)
Waste heat, pressurized water reactor, (Module 3): 21
Waste management, nuclear fuel, (Module 3): 20–22
Water heating systems, solar powered, (Module 4): 1, 4–7
Water hyacinth, (Module 2): 4
Water wheel, (Module 5): 2
Watt (W), (Module 1): 8, 34, (Module 4): 12
Watt-hours (Wh), (Module 4): 12, 33
Watts Bar nuclear plant, (Module 1): 12, (Module 3): 26, 29
Wave energy, (Module 1): 18, 20
Weather, (Module 2): 7
Willow trees, (Module 2): 5
Wind, origin of the, (Module 5): 8
Wind and Water Program (DOE), (Module 5): 6
Wind Application Centers (WACs), (Module 5): 7
Wind data, acquisition and use
    instruments, (Module 5): 12
    met towers, (Module 5): 11
    SCADA systems for, (Module 5): 26
    wind maps and charts, (Module 5): 11–12, 13
Wind Electrician Training Program, (Module 5): 4
Wind Energy by 2030 (DOE), (Module 5): 6
Wind energy equation, (Module 5): 8
Wind-energy sites, (Module 1): 13
Wind farms
    aesthetic appeal of, (Module 5): 1
    domestic examples, (Module 1): 14, (Module 5): 28, 29
    electric power produced by, (Module 1): 13–14
    locating, (Module 5): 1, 5, 27–28
    noise generated by, (Module 5): 1
    obstacles to development of, (Module 1): 14
    offshore, (Module 5): 5, 19–20
    post-oil crisis growth, (Module 5): 4
Wind industry
    career opportunities, (Module 5): 7
    organizations, (Module 5): 6
    present-day, (Module 5): 5–6
    research and development, (Module 5): 7
    standards, (Module 5): 6
Wind maps and charts, (Module 5): 11–12
Windmills, (Module 5): 2

Wind power
    advantages/disadvantages, (Module 5): 1
    available, determining
        averaging for, (Module 5): 11
        Betz Limit for, (Module 5): 15
        power density for, (Module 5): 14
    batteries for storing, (Module 1): 8, 10
    calculating, (Module 5): 12–15
    future of, (Module 5): 6–7
    global use of, (Module 5): 1, 5
    introduction, (Module 5): 1
    investment in, (Module 5): 1
    obstacles to development of, (Module 5): 1, 7
    origin and history of, (Module 5): 2–4
    overview, (Module 1): 13–14
    US capacity
        growth in, (Module 5): 1, 6–7
        percent of global total, (Module 5): 5
        requirements to increase, (Module 5): 6
    wind speed effect on, (Module 5): 9
Wind-powered systems
    grid-tied, (Module 5): 27–28
    off-grid, (Module 5): 26
    power storage, (Module 1): 10
    small applications, (Module 1): 10
Wind Powering America Program (WPA), (Module 5): 6–7, 12
Wind power technology education, (Module 5): 7
Wind rose, (Module 5): 12, 14, 36
Wind shear, (Module 5): 10–11, 36
Wind speed
    calculating, (Module 5): 8–9
    factors affecting, (Module 5): 8, 10
    height and, (Module 5): 9–10
    impact on power, (Module 5): 9
Wind technologies Market Report (DOE), (Module 5): 6
Wind technology, evolution of, (Module 5): 2–3
Wind turbine design
    basic types of, (Module 5): 15–16
    blade count in, (Module 5): 17
    blade size and configuration, (Module 5): 17–18
Wind turbines
    basic components, (Module 5): 16–17
    capacity, (Module 5): 20
    commercial, (Module 5): 17
    data collection, SCADA system, (Module 5): 25, 26
    gearbox-free, (Module 5): 22
    grid-connected, (Module 5): 4
    historically, (Module 5): 3
    largest, (Module 5): 17
    maintenance of, (Module 5): 20, 22
    manufacture of, (Module 5): 7, 17
    production units, (Module 1): 8
    research and development on, (Module 5): 17–18
    size of, (Module 5): 20
    transporting, (Module 5): 20. *See also* Horizontal-axis wind turbines (HAWTs); *specific components*
Wind vane, (Module 5): 17, 18
Wind velocity (V)
    calculating, (Module 5): 8–9
    height and, (Module 5): 9–10
Wood for biofuel, (Module 2): 1, 2
Wood pellets, (Module 2): 2, 12
Wood-related industries biomass use, (Module 2): 1
World War II, (Module 2): 8, 13, (Module 3): 27–28
World Wind Energy Association (WWEA), (Module 5): 6

WPA (Wind Powering America Program). *See* Wind
  Powering America Program (WPA)
WWEA (World Wind Energy Association). *See* World Wind
  Energy Association (WWEA)

Y
Yaw control, (Module 5): 22–23, 36
Yaw motors, (Module 5): 17, 18
Yellowcake, (Module 3): 5, 22
Yucca Mountain, (Module 3): 22

Z
Zero-energy district, (Module 1): 23, 34
Zero energy homes, (Module 4): 5